Jan Feiling

Optimization based on Non-Commutative Maps

Logos Verlag Berlin

Bibliografische Information der Deutschen Nationalbibliothek

Die Deutsche Nationalbibliothek verzeichnet diese Publikation in der
Deutschen Nationalbibliografie; detaillierte bibliografische Daten sind
im Internet über http://dnb.d-nb.de abrufbar.

D 93

ISBN 978-3-8325-5388-3

Logos Verlag Berlin GmbH
Georg-Knorr-Str. 4, Geb. 10,
D-12681 Berlin
Germany

Tel.:   +49 (0)30 / 42 85 10 90
Fax: +49 (0)30 / 42 85 10 92
http://www.logos-verlag.de

For my parents, Ina and Jürgen
and my brother, Tim.

Acknowledgements

This thesis is the result of my time as research and teaching assistant at the Institute of Systems Theory and Automatic Control (IST) at the University of Stuttgart. During this time, I was fortunate to meet many interesting and inspiring people who supported me in various ways and to whom I want to express my sincere gratitude.

First and foremost, I want to thank my advisor Prof. Dr.-Ing. Christian Ebenbauer for providing me the opportunity and the steady support to work on my doctoral degree. I am very grateful for the long and fruitful brainstorming sessions, discussions, and result presentations. Moreover, I enjoyed the freedom and I highly appreciated all the motivating words I always received, which empowered me to push beyond my boundaries. Eventually, the off-research topics in our discussions were very valuable for me and I am looking forward to keep this friendly relationship.

I also thank Prof. Mohamed-Ali Belabbas for supporting and challenging my research, as well as pointing me to the right directions.

I thank Dr. Christian Knörle, my mentor and supporter for my industrial and entrepreneurial career. The support and his many advises were and will be very valuable for me. I also thank the Forward31 team for pushing me on the final stage of this thesis.

Thanks to Roland Kurz, who motivated me to study engineering cybernetics.

A special thanks to my very good friend, former flatmate, and part-time office mate Fabian Pfitz. Even though we discussed more off-research topics and organized parties, this created a perfect balance to the work-intensive time.

Another special thanks goes to Dominik Widmann, a very close friend and fellow cybernetics student. Without him, especially without our sport sessions, I would not have enjoy cybernetics this much.

Thanks to Maximilian Schreiber, former colleague, very good friend, and flatmate, for pushing and motivating me on the final stage of this thesis, as well as discussing our future business ideas.

Thanks to all the team members of the IST. Although we did not have so much time together I really enjoyed this time, as well as, the time at various seminars and conferences.

I also want to thank all the students I supervised. Especially, Daniel Tremer, who become a very good friend and also my new neighbor in Berlin.

Thanks to my employer at this time, the Dr.-Ing. h.c. Porsche AG, as well as Dr.-Ing. Hans Georg Wahl and Matthias Worch who gave me the opportunity to focus on my research while providing me with the freedom to work on innovative topics in the Porsche ecosystem.

During the last stage of my thesis I met my girlfriend Saskia; I am thankful for her permanent support and her continuous flow of motivation to finish this thesis.

I am eternally grateful to the unconditional love and support of my family; my parents Ina und Jürgen and my brother Tim. You are my biggest source of motivation and always encouraged me—nothing of that would be possible without you.

Table of Contents

Abstract

In this thesis, we develop a novel class of discrete-time derivative-free optimization algorithms for unconstrained optimization problems. Our key idea is a new procedure to extract gradient information of an objective function using compositions of non-commutative maps. Those are defined by function evaluations and applied in such a way that gradient descent steps are approximated.

The procedure to construct a gradient approximation is based on two main ingredients: 1) a periodic exploration sequence that defines where the objective function is to be evaluated; 2) so-called gradient-generating functions that are composed with the objective function in such a way that an approximation of the gradient is obtained. Both ingredients can be characterized by a set of nonlinear equations. We propose a way to solve these equations, and we show how this leads to a derivative-free optimization algorithm with semi-global convergence properties.

The theoretical findings are supplemented with numerical results. A qualitative and quantitative simulation study is presented in which we investigate suitable design parameters, convergence speed, and gradient approximation errors of the proposed algorithm class. Further, we introduce various tuning rules such as variable step size schemes and adaptive exploration sequences. The algorithms we have developed are applied to challenging benchmarking problems and we compare them with other derivative-free optimization algorithms in their class. We validate that the presented algorithms are robust against noisy function evaluations, are able to deal with discontinuous objective functions, and potentially overcome local minima.

Deutsche Kurzfassung

In der vorliegenden Arbeit wird eine neue Klasse an zeitdiskreten ableitungsfreien Optimierungsalgorithmen für unbeschränkte Optimierungsprobleme entwickelt. Unsere Hauptidee ist ein neues Verfahren zur Bestimmung von Gradienteninformationen einer Funktion mittels Kompositionen nichtkommutativer Abbildungen. Diese sind durch Funktionsauswertungen definiert und ihre Anwendung führt zur Approximation von Gradientenabstiegsschritten.

Die Konstruktion dieses Verfahrens zur Gradientenapproximation basiert auf den folgenden zwei Bestandteilen: 1) periodische Explorationssequenzen, welche den nächsten Punkt zur Funktionsauswertung bestimmen; 2) Funktionen zur Gradientenerzeugung, bestehend aus Funktionsauswertungen der Optimierungsfunktion und einer analytischen Funktion, sodass ein Gradientenabstiegsschritt approximiert wird. Beide Bestandteile werden durch ein System von nichtlinearen Gleichungen charakterisiert. Wir stellen eine Lösung für diese Gleichungen vor und leiten damit einen ableitungsfreien Optimierungsalgorithmus mit semi-globalen Konvergenzeigenschaften her.

Die theoretischen Ergebnisse werden mit numerischen Resultaten ergänzt. Wir präsentieren eine qualitative und quantitative numerische Studie des Algorithmus, in der wir geeignete Designparameter sowie Konvergenzgeschwindigkeit und Gradientenapproximationsfehler untersuchen. Zur Performanzsteigerung wird eine abnehmende Schrittweitensteuerung und eine Adaptierung der Explorationssequenz vorgestellt. Außerdem werden numerische Experimente hinsichtlich verschiedener Benchmark-Probleme und Anwendungen diskutiert. Wir validieren, dass die vorgestellten Algorithmen ein robustes Verhalten gegenüber verrauschten Funktionsauswertungen haben, nicht stetiger Funktionen handhaben und potentiell lokale Minima überwinden können.

1

Introduction

1.1 Motivation and Background

Powerful optimization algorithms are key ingredients in science and engineering applications. Over the last decades, advances in machine learning, big-data-driven decision making, and real-time control methods have been accelerated by sophisticated optimization algorithms and the increasing computing power of microprocessors. Optimization in such applications is often very challenging, e.g. they are high-dimensional, non-convex, non-smooth, or of stochastic nature. Thus, improving existing optimization algorithms and developing novel algorithms is of central importance to master those challenges and therewith enhance technologies.

The need to solve increasingly complex optimization problems has particularly enabled the development of so-called *derivative-free or zeroth order optimization algorithms*, i.e., methods where no derivative information of the objective function is required—only function evaluations. This is especially appealing when the value of the objective function is obtained by simulations or other black box oracles, or where the calculation of the objective's derivative is computationally too expensive and only function evaluations are affordable. Such scenarios arise constantly in almost every area of modern technology and research: identifying and constructing the next drug against a disease, planning and scheduling the traffic flow in big cities, calculating the flight trajectory of a space mission, or developing an human-like artificial intelligence application, to name only a few potential applications.

The vast number of technology-driven applications and the increasing need for efficient optimization algorithms is emerging along with a growing number of publications in the area of derivative-free optimization algorithms. Historically, one of the earliest implementations of derivative-free algorithms was carried out on the von Neumann architecture-based computer *MANIAC 1*—an approximated solution of a six-dimensional non-linear least-squares problem calculated by utilizing derivative-free coordinate search (cf. Fermi (1952)). In the same year the well-known derivative-free optimization algorithm based on a gradient-approximation scheme by Kiefer, Wolfowitz, et al. (1952) was presented. Continuously, several extensions and improvements in the class of gradient approximations with so-called sample-averaging were developed, e.g in the work of Spall (1992) or Kushner and Clark (2012). The algorithms presented in this work are also derivative-free optimization algorithms based on gradient approximation ideas but are closely related to so-called *ex-*

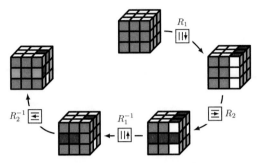

Figure 1.1. A figurative illustration of the mathematical concept of non-commutativity based on the magic cube. Rotations R_1 and R_2 and then their backward counterparts R_1^{-1} and R_2^{-1} are sequentially performed. Apparently, the initial configuration differs from the final configuration as a result of the non-commutativity of rotations R_1 and R_2.

tremum seeking control, a real-time (derivative-free) optimization method in control systems. Extremum seeking control can be traced back as far as 1922 to the work of Leblanc (1922), but only in the last two decades the field has regained new interest. Today, sophisticated tools for real-time optimization problems are available; see Krstić and Wang (2000), Teel and Popovic (2001), Guay and Zhang (2003), Tan, Moase, Manzie, Nešić, and Mareels (2010), and Benosman (2016), to mention just a few. One methodology in this field utilizes so-called *Lie brackets* from nonlinear geometric control (cf. Dürr, Stanković, Ebenbauer, and Johansson (2013) and Grushkovskaya, Zuyev, and Ebenbauer (2018)). Conceptually, this method is related to the approach we present in this thesis. However, as is quite common in control theory, extremum seeking schemes are stated as continuous time dynamical systems and not as discrete-time algorithmic optimization schemes.

In the present thesis, we develop a novel class of discrete-time, derivative-free optimization algorithms relying on gradient approximations based on *non-commutative maps*—inspired by the aforementioned Lie bracket approximation ideas in extremum seeking control systems. The introduced algorithm class has several interesting features. It shows robustness against noisy function evaluations, is able to deal with discontinuous objective functions, and potentially overcomes local minima. The main idea is to construct non-commutative maps with function evaluations to extract gradient information of the objective function. Conceptually speaking, two maps R_1, R_2 with a composition rule $\circ$ are commutative if their permuted compositions are identical (i.e., if $R_1 \circ R_2 = R_2 \circ R_1$). For example, multiplication of linear maps in the form of matrices is generally non-commutative, i.e., the result depends on the order in which they are multiplied. The way we utilize non-commutativity for optimization can be illustrated by the Magic Cube, known also as *Rubik's Cube* (Rubik (1975)), as depicted in Figure 1.1. Let R_i represent a rotation around an axis and R_i^{-1} its inverse counterpart; then it is obvious that the composition $R_1 \circ R_2 \circ R_1^{-1} \circ R_2^{-1}$

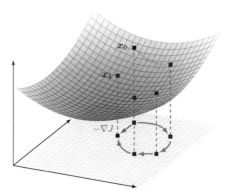

Figure 1.2. An illustration of the presented optimization algorithm class based on non-commutative maps. Effects of non-commutativity are utilized to approximate the negative gradient of the objective function and therewith the direction to a local minimum. The algorithm is initialized at x_0, and in four steps the algorithm performs an (approximated) negative gradient step to x_4.

does not commute as illustrated in Figure 1.1—the initial configuration is not equal to the final configuration. Figuratively speaking, the main idea and result of this work is the derivation of suitable mathematical definitions for maps R_i such that the difference between the initial and final configuration approximates the gradient of an objective function, as visualized in Figure 1.2. As a special case, two non-commutative maps and their inverse counterparts, similar to the Magic Cube illustration in Figure 1.1, are applied w.r.t. a point x_0 of an objective function. The gap between the first and last point, i.e., x_0 and x_4, is an approximation of the negative gradient at x_0 of the objective function (see Figure 1.1).

In a nutshell, this thesis is motivated by the idea to utilize the concept of non-commutative maps to approximate discrete-time gradient descent algorithms, i.e., designing a class of novel derivative-free optimization algorithms with convergence guarantees, various tuning parameters, and a scope of applications ranges from extremum seeking control problems to general (derivative-free) optimization problems.

1.2 Problem Statement

We consider unconstrained minimization problems of the form

$$\min_{x \in \mathbb{R}^n} J(x), \tag{1.1}$$

where only function evaluations of the objective $J : \mathbb{R}^n \to \mathbb{R}$ can be utilized to find a local minimum $x^* \in \mathbb{R}^n$ of J. The class of algorithms we propose is of the form

$$x_{k+1} = M_k^{\sqrt{h}}(x_k, J(x_k)), \quad k \geq 0 \tag{1.2}$$

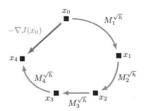

Figure 1.3. Principle of the algorithm class presented in this thesis. Composition of non-commutative maps as stated in (1.3) for $m = 4$ and $k = 0$ such that $x_4 - x_0 \approx -\nabla J(x_0)$, see (1.4).

where we call $M_k^{\sqrt{h}} : \mathbb{R}^n \times \mathbb{R} \to \mathbb{R}^n$ the *transition map* and $h \in \mathbb{R}_{>0}$ the step size. The main idea is to design the transition maps in such a way that for every $k \in \mathbb{N}_0$, the m-fold composition of these maps

$$x_{k+m} = \left(M_{k+m-1}^{\sqrt{h}} \circ \cdots \circ M_k^{\sqrt{h}} \right) (x_k, J(x_k)) \tag{1.3}$$

approximates a gradient descent step

$$x_{k+m} = x_k - h\nabla J(x_k) + \mathcal{O}(h^{3/2}), \tag{1.4}$$

as illuminated in Figure 1.3 (b). Conceptually, the "non-commutativity gap" between x_k and x_{k+m} represents the negative gradient of the objective function with an approximation error of order $h^{3/2}$. Thus, an approximated gradient-descent optimization procedure, well-known for example from finite-difference approximations (cf. Kiefer et al. (1952)). The term $\mathcal{O}(h^{3/2})$ in (1.4) represents a function $R : \mathbb{R}^n \times \mathbb{R} \times \mathbb{R}_{>0} \to \mathbb{R}^n$ such that for every compact convex set $\mathcal{X} \subseteq \mathbb{R}^n$ and $\mathcal{J} \subseteq \mathbb{R}$ there exist an $L \in \mathbb{R}_{>0}$ and $\bar{h} \in \mathbb{R}_{>0}$ such that for all $x_k \in \mathcal{X}$, $J(x_k) \in \mathcal{J}$, and $h \in [0, \bar{h}]$, $\|R(x_k, J(x_k); h^{3/2})\|_2 \le Lh^{3/2}$.

For the analysis of the algorithm (1.2) we impose the following assumptions.

Assumption 1. The objective function in the optimization problem (1.1) fulfills the following properties:

[A1] $J : \mathbb{R}^n \to \mathbb{R}$ is of class $C^2(\mathbb{R}^n, \mathbb{R})$, and transition maps $M_k^{\sqrt{h}} : \mathbb{R}^n \times \mathbb{R} \to \mathbb{R}^n$ for all $k \ge 0$ are of class $C^2(\mathbb{R}^n \times \mathbb{R}, \mathbb{R}^n)$.

[A2] $J : \mathbb{R}^n \to \mathbb{R}$ is radially unbounded, and there exists an $x^* \in \mathbb{R}^n$ such that $\nabla J(x)^\top (x - x^*) > 0$ for all $x \in \mathbb{R}^n \backslash \{x^*\}$.   •

We note that **[A2]** will be required only for the analysis of the convergence properties. The implementation of the algorithms, however, is not limited to the class of objective functions satisfying Assumption 1.

1.3 Contributions and Outline

In this thesis, we provide a novel class of derivative-free optimization algorithms that is based on non-commutative maps. In particular, we present a new procedure to extract gradient information from an objective function by constructing compositions of non-commutative maps. Those are defined by function evaluations and applied in such a way that gradient descent steps are approximated and semi-global convergence guarantees are given. We supplement our theoretical findings with numerical results. Therein, we provide several algorithm parameter studies and tuning rules, as well as the results of applying our algorithm to challenging benchmarking problems. The outline of the presented thesis is as follows:

- In Chapter 2, we first give an overview of related work. We highlight five main classes of derivative-free optimization algorithms and present established gradient approximation methods. Further, various applications where derivative-free optimization show good performance are discussed. Section 2.2 is dedicated to the concept of extremum seeking control that inspired the algorithms developed in this thesis. Finally, we present our main idea in Section 2.3. We derive our novel derivative-free algorithm class by two ingredients, namely the so-called *exploration sequence* and *gradient-generating functions*. The results of this chapter are based on Feiling, Zeller, and Ebenbauer (2018) and Feiling et al. (2019).

- In Chapter 3, we study the theoretical problem of the presented algorithm class. First, a general algorithmic scheme is proposed and its gradient approximation behavior is analyzed (Theorem 1). The two main ingredients are the generalized periodic exploration sequence that indicates where the objective is to be evaluated and a set of gradient-generating functions, which are composed with the objective function in such a way that an approximation of a gradient descent step is obtained. Based on that, the problem in Section 1.2 is approached by solving i) a quadratic system of equations and ii) a set of functional equations. This is related to the following two problems: 1) construction of the exploration sequence (Theorem 4) and 2) various cases of generating function pairs (Theorem 5 and Theorem 6), respectively. Given the solutions of i) and ii) by 1) and 2), the semi-global practical asymptotic convergence (Theorem 2) and, by some extension, semi-global asymptotic convergence (Theorem 3) to a local minimum of the objective function is proven. Eventually, we discuss the algorithm's design parameters and functions. This chapter's results are based on Feiling, Belabbas, and Ebenbauer (2020).

- In Chapter 4, we study the numerical problem of the presented algorithm class. A qualitative numerical study of the design parameters and functions is provided, as well as a quantitative numerical study with convergence speed and gradient approximation error as performance evaluation metric. Based on that, several performance tuning rules are discussed, namely variable and adaptive step sizes and an adaption of the exploration sequence. Finally, numerical experiments are carried out. In particular, a benchmarking study on challenging derivative-free optimization problems is presented, and we discuss potential applications of our algorithm class.

- In Chapter 5, we summarize the results of this thesis and conclude with an outlook of potential research directions.

Some technical background, notation, and preliminary lemmas are summarized in Appendix A. All technical proofs are gathered in Appendix B. Additional information on numerical results is provided in Appendix C and a step-by-step construction of the exploration sequence is presented in Appendix D. A list of nomenclature in this work is presented on page 123.

2

Related Work and Main Idea

In this chapter we establish the main idea of the presented algorithm class, which paves the way for the upcoming chapter. Before we present our main idea in Section 2.3, we give a brief overview of research in derivative-free optimization, gradient approximation schemes, and various applications in these fields (Section 2.1); because of the vast literature available in these research fields, we do not aim for a complete overview but refer to several survey and overview articles in the dedicated sections below. Since the presented algorithm class and its gradient approximation scheme is inspired by concepts from nonlinear geometric control and the continuous-time control method extremum seeking, we provide an introduction of these concepts in Section 2.2.

2.1 Related Work and Literature Review

Derivative-free optimization. This class of optimization algorithms requires no derivative of the objective function to find local minima (or a global minimum)—only function evaluations. Over the last decade, derivative-free optimization (as well as its naming twins black-box optimization, gradient-free optimization, optimization without derivatives, simulation-based optimization, and zeroth-order optimization) has been an active research field, that regained new interest after the early outstanding work in the sixties to eighties, e.g. in Fletcher (1965); Karmanov (1974); Matyas (1965); Nelder and Mead (1965); Polyak (1987); Rastrigin (1963); Rosenbrock (1960); ?. The acceleration in computational power in the last decades and the simplicity of applying derivative-free optimization methods were the main triggers for an increase in research publications in this field. Clearly, this class of algorithms is limited by accuracy, computational cost, or problem size, because of its strong correlation on the problem dimension and potential computationally expensive function evaluations. Nevertheless, the algorithms are known for their simple formulations and, thus, efficient implementation approaches. In this view, it is a class of algorithms that is very appealing for practical applications.

An overview of well-established, newly developed, and improved derivative-free optimization algorithms is presented in Conn, Scheinberg, and Vicente (2009), Audet and Hare (2017), and Rios and Sahinidis (2013) with a focus on software implementations for applications and industry problems. The algorithms can be clustered in five main categories:

- Direct search methods, first presented in Hooke and Jeeves (1961): only function evaluations are utilized, while no approximation of the gradient or the objective function is developed. Famous algorithms in this class are random and grid search (cf. Rastrigin (1963); J. Bergstra and Bengio (2012)), as well as the simplex optimization algorithm by Nelder and Mead (1965).

- Model-based methods, so-called surrogate or merit functions of the objective function serve as prediction models to calculate an update step of the algorithm by applying, for example, convex optimization principles and algorithms (cf. Boyd and Vandenberghe (2004)). A well-known method in this category is the trust region method (cf. Moré and Sorensen (1983); Conn, Gould, and Toint (2000)), of which various approaches to derive surrogate functions can be found in literature, e.g. polynomial models (cf. M. J. Powell (2003)), quadratic interpolation (cf. Winfield (1973)), or radial basis function interpolations (cf. Buhmann (2003)).

- Meta-heuristics, first mentioned in Fogel, Owens, and Walsh (1966): algorithms that mimic processes in natural selection, statistical mechanics, and population dynamics (cf. Holland et al. (1992)), for example, processes inspired by natural phenomena, such as grey wolf hunting behavior (cf. Mirjalili, Mirjalili, and Lewis (2014)) or the social behavior of humpback whales (cf. Mirjalili and Lewis (2016)). Famous and well-performing algorithms in this category are, for example, simulated annealing (cf. Kirkpatrick, Gelatt, and Vecchi (1983)), genetic algorithms (cf. Bonabeau, Dorigo, Marco, Theraulaz, and Théraulaz (1999)), or particle swarm optimization (cf. Eberhart and Kennedy (1995)).

- Bayesian optimization: this method is (often) based on Gaussian processes and sequential strategies motivated by statistical analysis (see, for example, Jones, Schonlau, and Welch (1998); Brochu, Cora, and De Freitas (2010); Shahriari, Swersky, Wang, Adams, and De Freitas (2015)). Conceptually, the objective function is treated as a random function with a prior distribution over the objective function beliefs. Those prior beliefs are updated via function evaluations to build a posterior distribution over the objective function beliefs. Based on that, the next search step is calculated via different versions of sampling criteria.

- (Stochastic) Approximation methods: derivatives, specifically first order information, i.e., gradients of the objective function, are approximated by so-called sample averaging of function evaluations. The first well-known algorithm in this class is the (scalar) method of Kiefer et al. (1952). Because this is the category of the presented optimization algorithm class of this thesis, we provide a more detailed discussion in the paragraph about gradient approximations below.

Note that many subcategories and various hybrid versions of algorithm classes in between the five categories, as stated above, exist. For a more detailed categorization we refer to Audet and Hare (2017) and Conn et al. (2009).

In our view, a further class of derivative-free optimization methods is extremum seeking control, which is derived as feedback controller for dynamical systems. Erroneously, this class is almost never mentioned in the classical derivative-free optimization literature, since it is derived and stated as a time-continuous system and not a discrete-time algorithmic scheme. Our work is inspired by extremum seeking control and its principles of nonlinear geometric control. Therefore, we dedicate Section 2.2 on this topic and provide a brief introduction to it.

As outlined above and in the mentioned references, many derivative-free algorithms exist, whereas every algorithm class and algorithm itself has advantages and disadvantages. On this account, benchmarking derivative-free algorithms (and optimization algorithms in general) is a common practice. State-of-the art and accepted benchmarking tests, i.e., problem classes and performance measurement definitions, for derivative-free optimization algorithms are presented in the work of Moré and Wild (2009) and Hansen, Finck, Ros, and Auger (2009). This topic is discussed in more detail in Section 4.5, where we carry out a benchmarking study for the proposed algorithm class on challenging objective functions to validate its practicality on non-convex, ill-conditioned, and discontinuous optimization problems.

Gradient approximation schemes. The first-order information of an objective function is extracted by utilizing function evaluations, producing a gradient approximation of the objective function. This is embedded into the gradient descent scheme

$$x_{k+1} = x_k - h\tilde{g}(J(x_k)), \tag{2.1}$$

where $\tilde{g} : \mathbb{R} \to \mathbb{R}^n$ is the gradient approximation of an objective $J : \mathbb{R}^n \to \mathbb{R}$, and $h \in \mathbb{R}_{>0}$ is the step size. Note that in the algorithm class presented in this thesis, (1.2) satisfies $\tilde{g}(J(x_k)) = \nabla J(x_k) + \mathcal{O}(\sqrt{h})$ as gradient approximation with an approximation error of order $\mathcal{O}(\sqrt{h})$ (see (1.4)).

In the following, three well-known gradient approximation schemes, which we revisit for benchmarking purposes in Chapter 4, are presented:

- Finite difference approximation: this scheme was introduced by Kiefer et al. (1952) for scalar objective functions, extended to the multi-variable case in Blum (1954), and applied to minimizing the sum-of-squares of non-linear functions by M. Powell (1965). The approximated gradient is calculated as

$$\tilde{g}(J(x)) = \left[\frac{J(x + \epsilon e_1) - J(x)}{\epsilon} \quad \frac{J(x + \epsilon e_2) - J(x)}{\epsilon} \quad \cdots \quad \frac{J(x + \epsilon e_n) - J(x)}{\epsilon} \right]^\top, \tag{2.2}$$

where the unit vectors $e_1, \ldots, e_n \in \mathbb{R}^n$ can be replaced by any set of vectors which form a basis of $\mathbb{R}^n$, and $\epsilon \in \mathbb{R}_{>0}$ can be scaled independently in each dimension if necessary.

- Random direction stochastic approximation: this method is an extension (cf. Kushner and Clark (2012)) of the finite difference approximation scheme. Namely, the unit vectors (or basis vectors of $\mathbb{R}^n$) are replaced by random direction vectors as follows:

$$\tilde{g}_k(J(x)) = \frac{J(x + \epsilon_k d_k) - J(x - \epsilon_k d_k)}{2\epsilon_k}. \tag{2.3}$$

The sequence $\{d_k\}$ consists of random direction vectors in $\mathbb{R}^n$, where in expectation it must hold $\mathbb{E}[d_k d_k^\top] = I$ and ϵ_k decreases with k. Note that $\tilde{g}_k$ depends on k.

- Simultaneous perturbation stochastic approximation: Spall (1992) extended and improved the convergence behavior of the finite difference and random direction stochastic approximation methods by the following multi-variable gradient approximation rule:

$$\tilde{g}_k(J(x)) = \frac{J(x + \epsilon_k d_k) - J(x - \epsilon_k d_k)}{2\epsilon_k} \left[(d_k e_1)^{-1} \quad (d_k e_2)^{-1} \quad \cdots \quad (d_k e_n)^{-1} \right]^\top. \quad (2.4)$$

The vector $d_k \in \mathbb{R}^n$ is constructed by n mutually independent mean-zero variables, and ϵ_k decreases with k. Additionally, Spall (1997) presents a one-measurement scheme for simultaneous perturbation stochastic approximation.

Note that all the presented gradient approximation methods (2.2) – (2.4) embedded into the gradient descent scheme (2.1) explore x_k in its ϵ_k-neighborhood. Therefore, we sometimes use the term *local exploration behavior* in the sequel. In contrast, algorithms such as random search and genetic algorithm tackle the exploration in a more global fashion by utilizing the entire defined search space.

Applications. Nowadays, in the presence of artificial intelligence and an increasing amount of new machine learning schemes, developers and users of those methods must face the problem of selecting suitable training parameters. The schemes are becoming more complex and computationally more expensive, and they involve more parameters to tune. A main application of derivative-free optimization algorithms is the tuning of these hyperparameters of artificial intelligence prediction models (cf. Bengio (2000); J. S. Bergstra, Bardenet, Bengio, and Kégl (2011); Li, Jamieson, DeSalvo, Rostamizadeh, and Talwalkar (2017)). Therein, a set of parameters is applied to the "learning procedure" of such prediction models, where the output in terms of a validation score (correct predictions of the validation data, not involved in the training data set) is the objective function to be optimized. Obviously, the goal is to maximize this validation score. A well-known hyperparameter toolbox is offered by "Google['s] internal service for performing black-box optimization that has become the de facto parameter tuning engine at Google. Google Vizier is used to optimize many of [Google's] machine learning models and other systems, and also provides core capabilities to Google's Cloud Machine Learning HyperTune subsystem", as stated in Golovin et al. (2017). The main applied algorithms in this toolbox are optimized and tailored for random and grid search algorithms, although whether more sophisticated derivative-free optimization algorithms are applied depends on the problem class. Another promising method for hyperparameter tuning are Bayesian optimization algorithms; see, for example, Snoek, Larochelle, and Adams (2012), Loshchilov and Hutter (2016), or Klein, Falkner, Bartels, Hennig, and Hutter (2016). Moreover, a summary of metaheuristic algorithms applied to hyperparameter optimization problems is presented in Mirjalili (2020).

Furthermore, derivative-free optimization algorithms are successfully applied to reinforcement learning problems, specifically to policy search problems (cf. Lehman, Chen,

Clune, and Stanley (2018); Salimans, Ho, Chen, Sidor, and Sutskever (2017)), where in the work of Mania, Guy, and Recht (2018), the best results for the MuJoCo benchmarking suite for model-based control (cf. Todorov, Erez, and Tassa (2012)) were achieved up to the publication date ot this thesis.

As previously mentioned, the concept of extremum seeking control provides a derivative-free optimization approach in a time-continuous system dynamical setting. Many practical applications involve real-time optimization tasks where the output of a dynamical system has to be optimized, e.g. operating a combustion engine at is optimal set point or maximizing the productivity of a bioreactor (cf. Ariyur and Krstic (2003)).

2.2 Extremum Seeking

The optimization algorithm class presented in this thesis is inspired from extremum seeking control. In this section we provide a brief introduction of this concept. It is a powerful class of model-free and real-time optimization methods in the field of control theory. A dynamical system is driven towards a steady state; that is, a (local) optimizer of an output map of the dynamical system, by utilizing time-periodic signals and output measurements. The first paper of extremum seeking control was probably that of Leblanc (1922). Since then, many extremum seeking approaches (see, e.g. Ariyur and Krstic (2003); Guay and Zhang (2003); Atta, Johansson, and Gustafsson (2015); Benosman (2016); Tan et al. (2010) for a detailed survey) have appeared in literature, especially with the regained interest initiated by the first rigorous proofs of stability for nonlinear continuous-time dynamical systems with extremum seeking feedback, based on averaging and singular perturbation theory (cf. Krstić and Wang (2000); Tan et al. (2010)). Extremum seeking for discrete-time systems and data-sampled-based extremum seeking has been considered, for example in Choi, Krstic, Ariyur, and Lee (2002); Stanković and Stipanović (2009); Teel and Popovic (2001); Khong, Nešić, Tan, and Manzie (2013); and Poveda and Teel (2017). Especially in Khong, Tan, Manzie, and Nešić (2015), gradient approximation schemes (2.2) – (2.4) are applied to a gradient descent scheme as in (2.1) and tailored to real-time optimization problems to optimize the output map of dynamical systems.

In recent years, a novel approach of continuous-time extremum seeking schemes has been designed by utilizing the non-commutativity of flows and the involved Lie bracket approximation methods. Based on these methods, a general framework to approximate gradients and to design extremum seeking systems for unconstrained, constrained, and distributed optimization and adaptive control problems has been developed; see, e.g. Dürr, Stanković, et al. (2013); Dürr, Zeng, and Ebenbauer (2013); Grushkovskaya et al. (2018); Scheinker and Krstić (2014); Michalowsky, Gharesifard, and Ebenbauer (2017); and Labar, Feiling, and Ebenbauer (2018).

This section is devoted to extremum seeking schemes that are approximated by so-called Lie bracket systems.

Definition 1. Let $f, g : \mathbb{R}^n \to \mathbb{R}^n$ be two continuously differentiable vector fields. Then,

$$[f, g] : z \mapsto \frac{\partial g}{\partial z}(z) f(z) - \frac{\partial f}{\partial z}(z) g(z) \tag{2.5}$$

is the Lie bracket of f, g (cf. Appendix A.2).   •

As presented in Dürr, Stanković, et al. (2013), the trajectory of scalar input-affine systems with state $x(t) \in \mathbb{R}$ of the form

$$\dot{x}(t) = f\big(J(x(t))\big) u(t) + g\big(J(x(t))\big) v(t) \tag{2.6}$$

with T-periodic inputs $u(t), v(t) \in \mathbb{R}$, where $T \in \mathbb{R}_{>0}$, are approximated by the trajectories of so-called Lie bracket systems

$$\dot{\bar{x}}(t) = \eta [f, g](\bar{x}(t)), \quad \bar{x}(t) \in \mathbb{R} \tag{2.7}$$

with

$$\eta = \frac{1}{2\pi} \int_0^T v(\theta) \int_0^\theta u(\tau) d\tau d\theta. \tag{2.8}$$

Specifically, under the assumption of twice continuously differentiable (scalar) vector fields $f, g : \mathbb{R} \to \mathbb{R}$ and the existence of uniformly bounded solutions of (2.7), the trajectory of (2.6) with initial state $x(0) = \bar{x}(0)$ are in a neighborhood of $\mathcal{O}(\sqrt{T})$ around the trajectory of (2.7), i.e., $|x(t) - \bar{x}(t)| \leq K\sqrt{T}$ for $K \in \mathbb{R}_{>0}$ (Theorem 1 in Dürr, Stanković, et al. (2013)).

Commonly, the periodic inputs in (2.6) are chosen as

$$u(t) = \sqrt{\omega} \sin(\omega t) \quad \text{and} \quad v(t) = \sqrt{\omega} \cos(\omega t) \quad \text{with} \quad \omega = \frac{2\pi}{T}, \tag{2.9}$$

where the vector fields are often picked as the following combinations

$$f(J(x)) = J(x), \qquad g(J(x)) = 1, \qquad \text{or} \tag{2.10}$$
$$f(J(x)) = \sin(J(x)), \qquad g(J(x)) = \cos(J(x)). \tag{2.11}$$

Combining (2.9) and (2.10) or (2.11) with (2.7) yields

$$\dot{\bar{x}}(t) = -\nabla J(\bar{x}(t)), \quad \bar{x}(t) \in \mathbb{R}; \tag{2.12}$$

thus, a gradient system. It is well known that a local strict minimum x^* of J is a stable equilibrium point of (2.12) (cf. Absil and Kurdyka (2006)), and therefore practical asymptotic stability of x^* of (2.6) can be proven (cf. Theorem 2 and Theorem 3 in Dürr, Stanković, et al. (2013)).

In terms of optimization theory, the concept of extremum seeking tackles (1.1) in a continuous-time fashion, where $J(x)$ acts as a static output map for the extremum seeking feedback as depicted in Figure 2.1 for (2.6) with (2.9) and (2.10). From a system design perspective, the vector fields f, g have to be chosen in a way that the Lie bracket between those represent the negative gradient of the objective function J. For detailed information on design and theory, including extension to the multi-variable version of (2.6), we refer to Dürr, Stanković, et al. (2013) and Grushkovskaya et al. (2018).

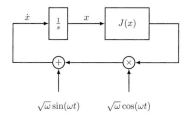

Figure 2.1. Extremum seeking control loop associated to (2.6) with (2.9) and (2.10).

2.3 Main Idea of Gradient Approximation

This section is partly taken from the work of Feiling, Zeller, and Ebenbauer (2018) and presents the main idea of this thesis. In terms of simplicity and based on a conceptual perspective, the problem statement is reduced to the scalar case in the remainder of this chapter, thus, $n = 1$ and $x \in \mathbb{R}$ in (1.1).

Let us approach the considered problem (1.1) with the following simple procedure: evaluate the objective function J at state x_k at iteration k, apply sinusoidal functions in a specific order, and update the state, such as

$$x_{k+1} = x_k + \sqrt{h} \begin{cases} \sin(J(x_k)) & \text{if } k \bmod (4) = 0 \\ \cos(J(x_k)) & \text{if } k \bmod (4) = 1 \\ -\sin(J(x_k)) & \text{if } k \bmod (4) = 2 \\ -\cos(J(x_k)) & \text{if } k \bmod (4) = 3 \end{cases} , \qquad (2.13)$$

where $k \in \mathbb{N}_0$ and $h \in \mathbb{R}_{>0}$ is the step size. Conceptually, this can be seen as perturbing the state x_k in a certain manner; therefore, exploring the space of the objective function around x_k to approximate a proper descent direction towards a local minimum. The first iterations of (2.13) are illustrated in Figure 2.2 with objective $J(x) = x^2$, global minimizer $x^* = 0$, and initial value $x_0 = 1$. As it is observed, the function value of J is decreasing in average and approaches the value $J(x^*) = 0$.

To understand this behavior from a mathematical perspective, *Taylor's Theorem* (T.T.) (see (Rudin, 1964, Theorem 5.15, p. 110) and Lemma 3 in Appendix A) is applied to the right-hand side of the evolution of (2.13) up to degree two. Then for the first four iteration

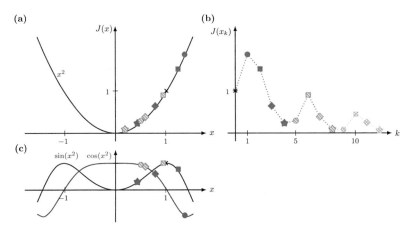

Figure 2.2. First 12 iterations steps of (2.13) with $h = 0.25$, objective $J(x) = x^2$, and initial value $x_0 = 1$. **Top Left (a)** Objective $J(x)$; **Top Right (b)** Periodical decreasing cost function value $J(x_k)$ w.r.t. k; **Bottom Left (c)** Sinusoidal functions with $J(x) = x^2$ as argument.

steps of (2.13), the expansions result in

$$x_1 = x_0 + \sqrt{h}\sin(J(x_0)) \tag{2.14}$$

$$x_2 = x_1 + \sqrt{h}\cos(J(x_1))$$
$$\overset{T.T.}{=} x_0 + \sqrt{h}\big(\sin(J(x_0)) + \cos(J(x_0))\big) - h\sin^2(J(x_0))\nabla J(x_0) + \mathcal{O}(h^{3/2}) \tag{2.15}$$

$$x_3 = x_2 - \sqrt{h}\sin(J(x_2))$$
$$\overset{T.T.}{=} x_0 + \sqrt{h}\cos(J(x_0))$$
$$- h\big(\sin^2(J(x_0)) + \cos^2(J(x_0)) + \sin(J(x_0))\cos(J(x_0))\big)\nabla J(x_0) + \mathcal{O}(h^{3/2}) \tag{2.16}$$

$$x_4 = x_3 - \sqrt{h}\cos(J(x_3))$$
$$\overset{T.T.}{=} x_0 - h\big(\sin^2(J(x_0)) + \cos^2(J(x_0))\big)\nabla J(x_0) + \mathcal{O}(h^{3/2})$$
$$= x_0 - h\nabla J(x_0) + \mathcal{O}(h^{3/2}). \tag{2.17}$$

It is worth mentioning that the above calculation is the discrete-time counterpart of a "calculation that everyone should do once in his[/her] life", as mentioned by R. W. Brockett (1976). As stated in (2.17), an approximated gradient descent step with a remainder of order $\mathcal{O}(h^{3/2})$ is performed every fourth iteration. Consequently, a descent is achieved as long as the gradient term dominates the remainder term, which is the case for h sufficiently small.

The algorithm of the form (2.13) is motivated by the continuous-time system (2.6). Therein,

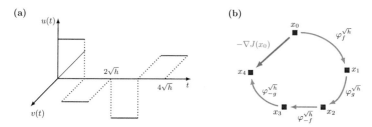

Figure 2.3. Inputs and flow maps of (2.6). **Left (a)** Rectangular-shaped periodic inputs $u(t), v(t)$ depicted for one period $4\sqrt{h}$; **Right (b)** Non-commutative flow maps of vector fields $\pm f, \pm g$.

the inputs $u(t)$ and $v(t)$ are selected as periodic rectangular-shaped signals, such as

$$
u(t) = \begin{cases} 1 & t \in [0, \sqrt{h}[\\ 0 & t \in [\sqrt{h}, 2\sqrt{h}[\\ -1 & t \in [2\sqrt{h}, 3\sqrt{h}[\\ 0 & t \in [3\sqrt{h}, 4\sqrt{h}[\end{cases}, \quad v(t) = \begin{cases} 0 & t \in [0, \sqrt{h}[\\ 1 & t \in [\sqrt{h}, 2\sqrt{h}[\\ 0 & t \in [2\sqrt{h}, 3\sqrt{h}[\\ -1 & t \in [3\sqrt{h}, 4\sqrt{h}[\end{cases}, \tag{2.18}
$$

depicted in Figure 2.3 (a). Then for small $h \in \mathbb{R}_{>0}$, similar to (2.14) – (2.17), and as presented in R. Brockett (2014), applying Taylor's Theorem (see (Rudin, 1964, Theorem 5.15, p. 110) and Lemma 3 in Appendix A) to the right hand side of (2.6) up to degree two for $t \in [0, 4\sqrt{h}[$ reveals

$$
x(4\sqrt{h}) = x(0) + h[f, g](x(0)) + \mathcal{O}(h^{3/2}). \tag{2.19}
$$

Choosing f, g in (2.19), as in (2.11), yields $[f, g](x) = -\nabla J(x)$ and

$$
x(4\sqrt{h}) = x(0) - h\nabla J(x(0)) + \mathcal{O}(h^{3/2}). \tag{2.20}
$$

Thus, with $x_k := x(4\sqrt{h})$, (2.20) and (2.17) are identical.

In principle, one could just numerically integrate (2.6) with vector fields f, g and inputs u, v as defined above to simulate the derivative-free discrete-time algorithm (2.20). However, rectangular-shaped inputs as shown in Figure 2.3 (a) often lead to bad numerical behavior in combination with standard numerical integration schemes. Thus, in continuous-time extremum seeking (cf. Dürr, Stanković, et al. (2013), Grushkovskaya et al. (2018)), the rectangular-shaped inputs are often replaced by sinusoidal inputs to avoid these problems (see (2.6) – (2.9)).

The key concept of the proposed derivative-free optimization algorithm class in this thesis is to directly design non-commutative maps instead of performing a numerical integration. This is motivated by our main idea to consider flow maps instead of vector fields. If we denote with $\varphi_\diamond^t(x_0) : \mathbb{R} \to \mathbb{R}$ the flow map of the vector fields $f, g : \mathbb{R} \to \mathbb{R}$ (in general

$f, g, \varphi_\diamond^t : \mathbb{R}^n \to \mathbb{R}^n)$ at time $t + t_0$ and initial condition $x(t_0) = x_0 \in \mathbb{R}$, where $\diamond = \{\pm f, \pm g\}$, then we can express (2.19) based on (2.6) and (2.18) for $t \in \,]0, 4\sqrt{h}[$ by

$$x(4\sqrt{h}) = \left(\varphi_{-g}^{\sqrt{h}} \circ \varphi_{-f}^{\sqrt{h}} \circ \varphi_{g}^{\sqrt{h}} \circ \varphi_{f}^{\sqrt{h}} \right)(x_0). \tag{2.21}$$

A visualization of these, in general non-commutative maps, is provided in Figure 2.3 (b) with the negative gradient at x_0 as a "non-commutative gap". Non-commutativity is related to Lie brackets, as in (2.19), given that they naturally arise when studying the commutativity of flow maps; they are an infinitesimal measure for the commutativity of maps. It is known (cf. Nijmeijer and Van der Schaft (1990)) that $x(4\sqrt{h}) = x(0)$ if and only if the flow maps $\varphi_f^{\sqrt{h}}, \varphi_g^{\sqrt{h}}$ commute, given that the flow is a bijection with $(\varphi_f^{\sqrt{h}})^{-1} = \varphi_{-f}^{\sqrt{h}}$ and thus $x(4\sqrt{h}) = x(0)$ if and only if $(\varphi_g^{\sqrt{h}} \circ \varphi_f^{\sqrt{h}})(x(0)) = (\varphi_f^{\sqrt{h}} \circ \varphi_g^{\sqrt{h}})(x(0))$, i.e., the Lie bracket is the zero vector.

The main goal of this thesis is to design flow maps and composition rules (input functions), similar to (2.18) and (2.21), such that the composition of maps approximates a gradient descent step as in (2.20). Thus, two challenges arise:

i) How to obtain these non-commutative maps?

ii) What are suitable discrete-time input functions?

i) Maps. We construct flow maps as in (2.21) by a scheme motivated from numerical integration methods, where we consider two approaches. First, we utilize one Euler-integration step. It does not lead to a Lie bracket approximation (cf. Altafini (2016)) as in (2.19), given that Euler-integration steps are a first-order method whereas Lie brackets are second-order effects. However, we show that an Euler-integration step still can be used for gradient approximations. Second, we utilize a so-called Heun-integration step, also known from the trapezoidal-integration method, that preserves properties like (2.19). Accordingly, for an ODE $\dot{x}(t) = s(J(x(t)))$ with $x(t) \in \mathbb{R}$ and a so-called *evaluation map* $s_\ell : \mathbb{R} \to \mathbb{R}$, we use the Euler- and Heun-integration-like steps defined as

$$E_\ell^{\sqrt{h}}(x_k, J(x_k)) := x_k + \sqrt{h} s_\ell(J(x_k)) \quad \text{and} \tag{2.22}$$

$$H_\ell^{\sqrt{h}}(x_k, J(x_k)) := x_k + \frac{\sqrt{h}}{2} \left(s_\ell(J(x_k)) + s_\ell(J(x_k + \sqrt{h} s_k(J(x_k)))) \right) \tag{2.23}$$

for approximating the flow map $\varphi_{s_\ell}^{\sqrt{h}}$. The maps (2.22) and (2.23), which we refer to as *transition maps*, are *single-point* and *two-point* maps, indicating the number of function evaluations per iteration.

ii) Inputs. Similar to the continuous-time periodic rectangular-shaped inputs in Figure 2.3 (a) and (2.18) we define so-called m-periodic exploration sequences $u_\ell, v_\ell \in \mathbb{R}$ for $\ell = 0, \ldots, m - 1$ such that the evaluation map s_ℓ is given by

$$s_\ell(J(x_k)) = f(J(x_k))u_\ell + g(J(x_k))v_\ell. \tag{2.24}$$

Note that we use the term exploration, given that s_ℓ is utilized to "explore" the neighborhood of the state x_k of the objective function J, similar to the gradient approximation methods presented in Section 2.1. As we will present in the next chapter, one of the main questions is how to construct $u_\ell, v_\ell \in \mathbb{R}^n$ for multi-variable optimization problems.

In this view, the algorithm in our motivating example (2.13) can be expressed by

$$
\begin{aligned}
x_{k+1} &= E_k^{\sqrt{h}}(x_k, J(x_k)) \\
&= x_k + \sqrt{h}(f(J(x_k))u_\ell + g(J(x_k))v_\ell)
\end{aligned}
\tag{2.25}
$$

with exploration sequences

$$
u_\ell = \begin{cases} 1 & \ell = 0 \\ 0 & \ell = 1 \\ -1 & \ell = 2 \\ 0 & \ell = 3 \\ u_{\ell-4} & \text{else} \end{cases}, \qquad
v_\ell = \begin{cases} 0 & \ell = 0 \\ 1 & \ell = 1 \\ 0 & \ell = 2 \\ -1 & \ell = 3 \\ v_{\ell-4} & \text{else} \end{cases}
\tag{2.26}
$$

and functions f, g as in (2.11). Sequentially applying the Taylor expansion as in (2.14) – (2.17) again reveals

$$
\begin{aligned}
x_{k+4} &= \left(E_{k+3}^{\sqrt{h}} \circ E_{k+2}^{\sqrt{h}} \circ E_{k+1}^{\sqrt{h}} \circ E_k^{\sqrt{h}} \right)(x_k) \\
&= x_k - h\nabla J(x_k) + \mathcal{O}(h^{3/2}),
\end{aligned}
\tag{2.27}
$$

as demanded. Summarizing, (2.22) and (2.23), together with (2.24) and exploration sequences u_ℓ, v_ℓ, define an algorithm class as proposed in (1.2), i.e., $M_k^{\sqrt{h}}(x_k, J(x_k))$ is given by (2.22) or (2.23), accordingly. In the next chapter we provide a generalization of the non-commutative maps (2.22) and (2.23), as well as of the evaluation map $s_\ell : \mathbb{R} \to \mathbb{R}^n$ in (2.24) and $u_\ell, v_\ell \in \mathbb{R}^n$ for multi-variable optimization problems.

3

Derivative-Free Optimization based on Non-Commutative Maps

This chapter is partly taken from the work of Feiling et al. (2020) and studies the theoretical problem of this thesis and the presented algorithm class. First, we introduce a general parameterized structure of the non-commutative maps such that a gradient descent step is approximated (Section 3.1). Second, we analyze the convergence behavior of the presented algorithm class (Section 3.2). In the two subsequent sections we derive the components to construct general non-commutative maps for our algorithm class, split into the design of exploration sequences (Section 3.3) and gradient-generating functions (Section 3.4). The last part of this chapter is dedicated to the algorithm design parameters with several examples (Section 3.5), as well as a summary and outlook part (Section 3.6).

3.1 The Algorithmic Scheme

For the proposed algorithms of the form

$$x_{k+1} = M_k^{\sqrt{h}}(x_k, J(x_k)), \quad k \geq 0, \tag{3.1}$$

with *transition maps* $M_k^{\sqrt{h}} : \mathbb{R}^n \times \mathbb{R} \to \mathbb{R}^n$, we impose the structure based on (2.22) and (2.23) in Section 2.3 as

$$M_k^{\sqrt{h}}(x_k, J(x_k)) = x_k + \sqrt{h}\alpha_1 s_k\Big(J(x_k)\Big) + \sqrt{h}\alpha_2 s_k\Big(J\big(x_k + \sqrt{h}s_k(J(x_k))\big)\Big)$$

$$s_\ell(J(x_k)) = f(J(x_k))u_\ell + g(J(x_k))v_\ell \tag{3.2}$$

with parameters $\alpha_1, \alpha_2 \in \mathbb{R}$ where $\alpha_1 + \alpha_2 \neq 0$. We call $s_\ell : \mathbb{R} \to \mathbb{R}^n$ the *evaluation map*, $f, g : \mathbb{R} \to \mathbb{R}$ *gradient-generating functions*, and $u_\ell, v_\ell \in \mathbb{R}^n$ m-periodic exploration sequences. Note that for $\alpha_2 \neq 0$, only two evaluations of J per iteration are necessary. In this view, we say that the algorithms (3.1) with (3.2) and $\alpha_2 = 0$ have a *single-point characteristic*, otherwise a *two-point characteristic*. Note that we use the term gradient-generating functions instead of generating vector fields (cf. Dürr, Stanković, et al. (2013); Grushkovskaya et al. (2018)) because f, g are scalar functions. In contrast, we define multi-dimensional periodic exploration sequences u_ℓ, v_ℓ. We elaborate on the choice of this structure for the algorithm in the sequel. The main goal of this chapter is to characterize and design

1. m-periodic exploration sequences u_ℓ, v_ℓ and

2. gradient-generating functions f and g

such that (3.1) with transition map (3.2) yields a gradient descent step as in (1.4). Note that Assumption **[A1]** implies that f, g are of the class $C^2(\mathbb{R}, \mathbb{R})$. Our first result in this chapter restates the problem, as introduced in Section 1.2, in terms of solving a system of nonlinear equations. Because the sequences u_ℓ and v_ℓ are m-periodic, it suffices to determine the steps $k = 0, \ldots, m - 1$ in (3.2), i.e., one period of the exploration sequences.

Theorem 1. Let **[A1]** hold. Then m evolutionary steps of (3.1) with transition map (3.2) is given by

$$x_{k+m} = x_k + \sqrt{h}(\alpha_1 + \alpha_2)Y\Big(f\big(J(x_k)\big), g\big(J(x_k)\big)\Big)W\mathbb{1}$$
$$+ h\tilde{Y}\Big(f\big(J(x_k)\big), g\big(J(x_k)\big)\Big)T(W)Y\Big(f\big(J(x_k)\big), g\big(J(x_k)\big)\Big)^\top \nabla J(x_k) + \mathcal{O}(h^{3/2}). \quad (3.3)$$

Here, $W = [w_k \ w_{k+1} \ \cdots \ w_{k+m-1}] \in \mathbb{R}^{2n \times m}$ is the exploration sequence matrix with $w_i = [u_i^\top \ v_i^\top]^\top \in \mathbb{R}^{2n}$, and $T(W) \in \mathbb{R}^{2n \times 2n}$ is given by

$$T(W) := \sum_{i=0}^{m-1} \left(\alpha_2 w_i w_i^\top + (\alpha_1 + \alpha_2)^2 \sum_{j=0}^{i-1} w_i w_j^\top \right). \quad (3.4)$$

Furthermore,

$$Y\big(f(z), g(z)\big) := \big[f(z)I \ \ g(z)I\big] \in \mathbb{R}^{n \times 2n} \text{ and} \quad (3.5)$$
$$\tilde{Y}\big(f(z), g(z)\big) := \left[\frac{\partial f}{\partial z}(z)I \ \ \frac{\partial g}{\partial z}(z)I\right] = \frac{\partial Y}{\partial z}\big(f(z), g(z)\big) \in \mathbb{R}^{n \times 2n} \quad (3.6)$$

are defined. ●

The proof of Theorem 1 is given in Appendix B.1. If there exist m-periodic exploration sequences $\{w_\ell\}_{\ell=0}^{m-1}$ (equivalently, an exploration sequence matrix W) and gradient-generating functions f and g such that

$$W\mathbb{1} = 0 \quad (3.7)$$
$$\tilde{Y}\big(f(z), g(z)\big)T(W)Y\big(f(z), g(z)\big)^\top = -I, \ \forall z \in \mathbb{R} \quad (3.8)$$

are satisfied, then (1.4) holds. Thus, this system of nonlinear ordinary differential equations (w.r.t. $f(z)$ or $g(z)$) with unknown coefficients is key in designing the algorithm. The idea to solve this highly under-determined system of equations is now to proceed in two steps:

Step 1) For a class of normal (skew-symmetric) matrices T_d, we construct exploration sequence matrices W such that (3.7) and $T(W) = T_d$ hold.

Step 2) We characterize gradient-generating functions f, g and normal (skew-symmetric) matrices T_d such that (3.8) hold.

These two steps are presented in Section 3.3 and Section 3.4, respectively. Beforehand, we start with a remark and example on $T(W)$ and the convergence result of the proposed algorithm in the next section.

Remark 1. To get a sense of equation (3.8) and the role of $T(W)$, partition $T(W)$ as

$$T(W) = \begin{bmatrix} T_{11}(W) & T_{12}(W) \\ T_{21}(W) & T_{22}(W) \end{bmatrix}, \tag{3.9}$$

with $T_{11}(W), T_{12}(W), T_{21}(W), T_{22}(W) \in \mathbb{R}^{n \times n}$. Note that $T_{11}(W)$ is defined by $\{u_\ell\}_{\ell=0}^{m-1}$, $T_{22}(W)$ by $\{v_\ell\}_{\ell=0}^{m-1}$, and $T_{12}(W)$ and $T_{21}(W)$ by both $\{u_\ell\}_{\ell=0}^{m-1}$ and $\{v_\ell\}_{\ell=0}^{m-1}$. Then (3.8) with (3.9) yields

$$\frac{\partial f}{\partial z}(z)f(z)T_{11}(W) + \frac{\partial f}{\partial z}(z)g(z)T_{12}(W)$$
$$+ \frac{\partial g}{\partial z}(z)f(z)T_{21}(W) + \frac{\partial g}{\partial z}(z)g(z)T_{22}(W) = -I. \tag{3.10}$$

A geometric interpretation of $T(W)$ is discussed in Section 3.3. ●

Example 1. A coordinate-wise descent with $m = 4n$ (cf. Lemma 1 and Lemma 2 in Feiling, Zeller, and Ebenbauer (2018)) is achieved by choosing the exploration sequences as

$$u_\ell = \bar{u}_\ell e_i, \quad v_\ell = \bar{v}_\ell e_i \text{ with } i = \lfloor \ell/4 \rfloor \bmod(n) + 1$$

$$\bar{u}_\ell = \begin{cases} 1 & \ell = 0 \\ 0 & \ell = 1 \\ -1 & \ell = 2, \\ 0 & \ell = 3 \\ \bar{u}_{\ell-4} & \text{else} \end{cases} \qquad \bar{v}_\ell = \begin{cases} 0 & \ell = 0 \\ 1 & \ell = 1 \\ 0 & \ell = 2 \\ -1 & \ell = 3 \\ \bar{v}_{\ell-4} & \text{else} \end{cases}. \tag{3.11}$$

Note that this also corresponds to (2.26). This leads for $n = 2$ to the exploration sequence matrix

$$W = \begin{bmatrix} 1 & 0 & -1 & 0 & 0 & 0 & 0 & 0 \\ 0 & 0 & 0 & 0 & 0 & 0 & -1 & 0 \\ 0 & 1 & 0 & -1 & 0 & 0 & 0 & 0 \\ 0 & 0 & 0 & 0 & 0 & 1 & 0 & -1 \end{bmatrix}, \tag{3.12}$$

where $W\mathbb{1} = 0$ is satisfied. Therefore, the terms of order $\sqrt{h}$ in (3.3) vanish. Then with Theorem 1, i.e., (3.3) – (3.6), the matrix $T(W) \in \mathbb{R}^{4\times4}$ is for $[\alpha_1\ \alpha_2] = [1\ 0]$ (single-point characteristic) given by

$$T(W) = \begin{bmatrix} -I & -I \\ I & -I \end{bmatrix}, \tag{3.13}$$

and the evolution of x_k yields

$$x_{k+m} = x_k + h\Big\{ ([f,g](J(x_k)) - \frac{1}{2}\frac{\partial(f^2+g^2)}{\partial J}(J(x_k)) \Big\}\nabla J(x_k) + \mathcal{O}(h^{3/2}). \tag{3.14}$$

On the same lines, it holds for $[\alpha_1\ \alpha_2] = [1/2\ 1/2]$ (two-point characteristic)

$$T(W) = \begin{bmatrix} 0 & -I \\ I & 0 \end{bmatrix}; \tag{3.15}$$

thus,

$$x_{k+m} = x_k + h\Big\{ ([f,g](J(x_k)) \Big\}\nabla J(x_k) + \mathcal{O}(h^{3/2}). \tag{3.16}$$

Note that (3.14) and (3.16) hold for $n \in \mathbb{N}$ arbitrary. A simple calculation shows that the term in curly brackets in (3.14) and (3.16) is identical to -1 for $f(z) = \sin(z)$ and $g(z) = \cos(z)$; thus, a gradient descent step as in (1.4) is performed.   •

3.2 Convergence

In this section we assume that (3.7) and (3.8) are satisfied, thus, there exists a triple (W, f, g) such that (3.3) leads to a gradient descent step as in (1.4):

$$x_{k+m} = x_k - h\nabla J(x_k) + \mathcal{O}(h^{3/2}). \tag{3.17}$$

Then semi-global practical asymptotic convergence (cf. Moreau and Aeyels (2000); Khalil (2002); Vidyasagar (2002)) to the optimizer x^* can be established:

Theorem 2. Let **[A1]** and **[A2]** hold. Assume that there exist gradient-generating functions f, g and an exploration sequence matrix W such that (3.7) and (3.8) are satisfied. Then for all $\delta_1, \delta_2 \in \mathbb{R}_{>0}$ there exist an $h^* \in \mathbb{R}_{>0}$ and $N \in \mathbb{N}_0$, such that for all $h \in\ ]0, h^*[$ and $x_0 \in \mathcal{U}_{\delta_1}(x^*)$, it holds $x_k \in \mathcal{U}_{\delta_2}(x^*)$ for all $k \geq N$.   •

The proof of Theorem 2 is given in Appendix B.2. Given that the step size h is constant in Theorem 2, the remainder term of order $\mathcal{O}(h^{3/2})$ is almost never vanishing. Applying a variable decreasing step size h_k, that is constant over a period of length m, such as

$$h_\ell = h_{\ell+1} = \cdots = h_{\ell+m-1} \text{ for } \ell = pm, \ p = 0, 1, \ldots, \text{ with} \tag{3.18a}$$

$$\sum_{p=0}^{\infty} h_{pm} = \infty, \quad \sum_{p=0}^{\infty} h_{pm}^2 < \infty, \tag{3.18b}$$

for example, $h_k = (\lfloor k/m \rfloor + 1)^{-1}$ (cf. Proposition 1 in Spall (1992)) leads to semi-global asymptotic convergence.

Theorem 3. Let **[A1]** and **[A2]** hold. Assume that there exist gradient-generating functions f, g and an exploration sequence matrix W such that (3.7) and (3.8) are satisfied. Moreover, let h_k satisfy (3.18) for all $k \in \mathbb{N}_0$. Then for all $\delta_1 \in \mathbb{R}_{>0}$ there exists an $h_0^* \in \mathbb{R}_{>0}$, such that for all $h_0 \in \,]0, h_0^*[$ and $x_0 \in \mathcal{U}_{\delta_1}(x^*)$, it holds $x_k \to x^*$ for $k \to \infty$. •

The proof of Theorem 3 is given in Appendix B.3. Note that the requirement of periodically m constant steps preserves the $\mathcal{O}(\sqrt{h})$-order terms in (3.3) (cf. proof of Lemma 6).

3.3 Exploration Sequence

In this section we characterize the conditions under which there exists an exploration sequence matrix W for a given $T_d \in \mathbb{R}^{2n \times 2n}$ such that $T(W) = T_d$ together with $W\mathbb{1} = 0$ are satisfied, thus, addressing Step 1) as stated in Section 3.1. The next lemma represents $T(W)$, given in (3.4), in combination with (3.7) in a more compact form.

Lemma 1. Consider (3.4) and suppose the exploration sequence matrix $W \in \mathbb{R}^{2n \times m}$ satisfies (3.7). Then $T(W)$ can be expressed as

$$T(W) = WPW^{\top}, \tag{3.19}$$

with $P \in \mathbb{R}^{m \times m}$ defined as

$$P = \begin{bmatrix} c_1 & c_2 & \cdots & c_2 & 0 \\ \alpha_2 & \ddots & \ddots & \vdots & \vdots \\ \vdots & \ddots & \ddots & c_2 & \vdots \\ \alpha_2 & \cdots & \alpha_2 & c_1 & 0 \\ 0 & \cdots & \cdots & 0 & 0 \end{bmatrix}, \tag{3.20}$$

where $c_1 = 2\alpha_2 - (\alpha_1 + \alpha_2)^2$, $c_2 = \alpha_2 - (\alpha_1 + \alpha_2)^2$, and α_1, α_2 defined in (3.2). •

The proof of Lemma 1 is given in Appendix B.4. Consequently, when proceeding according to Step 1) and Step 2) as described in Section 3.1, the key equations for designing exploration sequences are

$$\begin{aligned} WPW^{\top} &= T_d \\ W\mathbb{1} &= 0 \, . \end{aligned} \tag{3.21}$$

The following theorem, which provides a constructive design of the exploration sequence matrix W, is of central importance. It also provides structural insights in terms of obtaining lower bounds on the sequence length (period) m, suitable choices of the parameters α_1, α_2, and admissible structures for the desired target matrices T_d.

Theorem 4. Given α_1, α_2 and $T_d \in \mathbb{R}^{2n \times 2n}$. Suppose that either T_d is normal, $(2\alpha_2 - (\alpha_1 + \alpha_2)^2)(T_d + T_d^\top)$ is positive definite, and Conjecture 1 (see below) is satisfied or that T_d is skew-symmetric with $2\alpha_2 - (\alpha_1 + \alpha_2)^2 = 0$. Then there exists a $m \geq \text{rk}(T_d) + 1$ such that $W \in \mathbb{R}^{2n \times m}$ satisfies the system of equations (3.21).   •

Conjecture 1. Let the skew-symmetric matrix $C(m) \in \mathbb{R}^{m \times m}$ be defined as

$$C(m) := A(m) + \epsilon(m + 1)B(m), \tag{3.22}$$

where

$$A(m) := \begin{bmatrix} 0 & 1 & \cdots & 1 \\ -1 & \ddots & \ddots & \vdots \\ \vdots & \ddots & \ddots & 1 \\ -1 & \cdots & -1 & 0 \end{bmatrix} \in \mathbb{R}^{m \times m}, \tag{3.23}$$

$$B(m) := \mathbb{1}[0\ 2\ 4\ \ldots\ 2(m-1)] - (\mathbb{1}[0\ 2\ 4\ \ldots\ 2(m-1)])^\top \in \mathbb{R}^{m \times m} \tag{3.24}$$

with $\epsilon(m) = (m-1)^{-1}(1 - m^{-1/2})$. Then for any $m \geq 2$, $C(m)$ and $C(m+1)$ satisfy the eigenvalue interlacing property (see Lemma 4 in Appendix A.2).

$$\omega_k^{m+1} > \omega_k^m > \omega_{k+1}^{m+1} \geq 0, \tag{3.25}$$

for $k = 1, \ldots, \lfloor m/2 \rfloor$, where $\pm \omega_k^m i$ are the eigenvalues of $C(m)$, with $\{\omega_k^m\}_{k=1}^{\lfloor m/2 \rfloor}$ sorted in non-decreasing order in k for m fixed.   •

Remark 2. We verified numerically that Conjecture 1 is always true (up to $m = 10000$, see Appendix A.3.1), but a proof is still lacking. Further notice, that $C(m)$ is part of

$$\begin{aligned} \bar{P}(m) &= \left(P - \epsilon(m)(\mathbb{1}\mathbb{1}^\top P + P\mathbb{1}\mathbb{1}^\top) + \epsilon^2(m)\mathbb{1}\mathbb{1}^\top P\mathbb{1}\mathbb{1}^\top\right)_{1:m-1} \\ &= \left(\frac{1}{2}(\alpha_1 + \alpha_2)^2 - \alpha_2\right)I + \frac{1}{2}(\alpha_1 + \alpha_2)^2 C(m-1) \end{aligned} \tag{3.26}$$

with P in (3.20) and $\epsilon(m)$ defined in Conjecture 1. Note that the interlacing eigenvalue property (3.25) holds also for $\bar{P}(m)$, because it is arranged by a scaled unit matrix and the skew-symmetric matrix $C(m-1)$. The interlacing property of $\bar{P}(m)$ is utilized in the proof of Theorem 4.   •

Corollary 1. If $2\alpha_2 - (\alpha_1 + \alpha_2)^2 = 0$ and T_d skew-symmetric, then there always exists a $W \in \mathbb{R}^{2n \times m}$ with $m = \text{rk}(T_d) + 1$.   •

The proof of Theorem 4 is constructive and presented in Appendix B.5, where Conjecture 1 particularly enters in (B.29). The role of Conjecture 1 in Theorem 4 is discussed in Remark 2 above and in Example 4 and Example 5 (p. 32f). Moreover, a step-by-step construction of W for a given T_d is provided in Appendix D and applied in Example 2 – Example 7 (p. 31ff). Note that Corollary 1 is directly derived by the proof of Theorem 4.

Geometric Interpretation of Exploration Sequences

In this subsection we provide a geometric interpretation and a visualization of the exploration sequence matrix W and $T(W)$, which is of interest on its own. The values of the components of $T(W)$ in (3.4) for $[\alpha_1 \ \alpha_2] = [1/2 \ 1/2]$, i.e.,

$$T(W)_{pq} = \sum_{i=0}^{m-1} \sum_{j=0}^{i-1} \frac{1}{2} e_p^\top w_i w_i^\top e_q + e_p^\top w_i w_j^\top e_q, \tag{3.27}$$

with $p = 1, \ldots, 2n$, where the index pq specifies the element of $T(W)$ in the p-th row and q-th column, can be interpreted as the projected areas on the $e_p - e_q$-plane spanned by the exploration sequences $\{e_p^\top w_\ell\}_{\ell=0}^{m-1}$, $\{e_q^\top w_\ell\}_{\ell=0}^{m-1}$, with the zero vector as initial and final value. Interestingly, the net area $A_{pq} \in \mathbb{R}$ of an n-sided polygon in the $x_p - y_q$-plane with corner points $(x_{p,i}, y_{q,i}) \in \mathbb{R}^2$, $i = 0 \ldots n - 1$ and $p = 1, \ldots, 2n$, known as the *Shoelace* or *Gauss* area formula (cf. Braden (1986)), is obtained as a special case of Green's Theorem and is given by

$$A_{pq} = \frac{1}{2} \sum_{i=0}^{n-1} \left(x_{p,i+1} y_{q,i} - x_{p,i} y_{q,i+1} \right) \tag{3.28}$$

with $x_{p,0} = x_{p,n} = 0$ and $y_{q,0} = y_{q,n} = 0$. The relation between $T(W)_{pq}$ in (3.27) and A_{pq} in (3.28) is stated in the next lemma.

Lemma 2. Let W such that $W\mathbb{1} = 0$ and $[\alpha_1 \ \alpha_2] = [1/2 \ 1/2]$ be given. Then the value in the p-th row and q-th column of $T(W)$ in (3.3) is equivalent to the net area of the n-sided polygon, i.e. (3.28), in the $x_p - y_q$-plane with corner points

$$x_{p,i} = \sum_{k=0}^{i-1} e_p^\top w_k, \quad y_{i,q} = \sum_{k=0}^{i-1} e_q^\top w_k \tag{3.29}$$

where $x_{p,0} = 0$ and $y_{q,0} = 0$ for all $p, q = 1, \ldots, 2n$. •

The proof of Lemma 2 is given in Appendix B.6. Moreover, an area generation for some $w_k = [u_k \ v_k]^\top \in \mathbb{R}^2$, $k = 0, \ldots, 5$ with $\sum_{k=0}^{5} w_k = 0$ is illustrated in Figure 3.1. This geometric interpretation is not surprising in light of the relation to nonholonomic integrator problems and the area generating rule appearing in the study of nonholonomic systems (see Bloch (2003)). In particular, (3.28) with (3.29) represent a (double) iterated sum over the exploration sequence. In the continuous-time setting, this would correspond to double iterated integrals (or in general k-fold iterated integrals, which are called the signature of a path), which play an important role in nonholonomic control systems. Note that for $[\alpha_1 \ \alpha_2] \neq [1/2 \ 1/2]$, Lemma 2 does not hold. However, we believe it is related to some kind of weighted areas. An illustration of this geometric interpretation for the exploration sequences in (3.45) is depicted in Figure 3.2. In Section 4.2 we elaborate more on the shape of the generated areas. We will identify that the singular values of the exploration sequence matrix W play a crucial role on the shape of the generated areas, as well as on the exploration sequence length m.

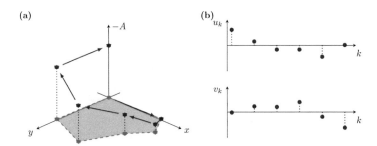

Figure 3.1. Area generation in the $x - y$-plane with $w_k = [u_k \ v_k]^\top$ for $k = 1, \ldots, 5$, as stated in Lemma 2. **Left (a)** Area in $x - y$-plane with the net area value as in (3.28) on the z-axis; **Right (b)** Sequences u_k and v_k, where $\sum_{k=0}^{5} u_k = 0$ and $\sum_{k=0}^{5} v_k = 0$.

In the following we point out two relations of the exploration sequence construction procedure to other research fields. This part does not directly contribute to the algorithm design but might be of interest for further extensions in the design of the exploration sequence.

Relation to Nonholonomic Integrator Problems

It is worthwhile to point out an interesting connection between the equations in (3.21) and nonlinear control theory, specifically the controlability of the nonholonomic integrator. Supposing $\{w_\ell\}_{\ell=0}^{m-1}$ is a solution of (3.21), it can be verified by direct calculations (see also proof of Lemma 1 in Appendix B.4) that it is also a solution of the two-point boundary value problem

$$y_0 = 0, \ Z_0 = 0, \ y_m = 0, \ Z_m = T_d$$
$$y_{k+1} = y_k + w_k \tag{3.30}$$
$$Z_{k+1} = Z_k + (\alpha_1 + \alpha_2)^2 w_k y_k^\top + \alpha_2 w_k w_k^\top$$

with $k = 0, \ldots, m - 1$, states $y_k \in \mathbb{R}^{2n}$, $Z_k \in \mathbb{R}^{2n \times 2n}$, input $w_k \in \mathbb{R}^{2n}$, and vice versa. In particular, with $W\mathbb{1} = 0$, i.e., $w_{m-1} = -\sum_{i=0}^{m-2} w_i$, (3.30) translates into

$$y_0 = 0, \ Z_0 = 0, \ Z_{m-1} = T_d$$
$$y_{k+1} = y_k + w_k$$
$$Z_{k+1} = Z_k + \alpha_2 w_k y_k^\top + (\alpha_2 - (\alpha_1 + \alpha_2)^2) y_k w_k^\top . \tag{3.31}$$
$$+ (2\alpha_2 - (\alpha_1 + \alpha_2)^2) w_k w_k^\top$$

Considering now the case $[\alpha_1 \ \alpha_2] = [1/2 \ 1/2]$, it shows that (3.31) is the state-transition of the generalized discrete-time nonholonomic integrator (cf. Altafini (2016)) with given initial

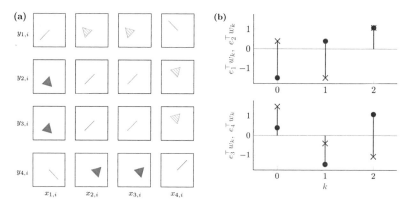

Figure 3.2. Exploration sequence and its generating area effects based on the exploration sequence matrix W in (3.45). **Left (a)** Generated areas with respect to the coordinate system $x_{p,i}$, $y_{q,i}$ for $p, q = 1, \ldots, 2n$ given by (3.29). For $T(W) \in \mathbb{R}^{4 \times 4}$ it holds $T(W)_{1,2} = T(W)_{1,3} = T(W)_{2,4} = T(W)_{3,4} = -1$, $T(W)_{2,1} = T(W)_{3,1} = T(W)_{4,2} = T(W)_{4,3} = 1$ and the rest 0. The filled green areas (■) have an area surface value of 1, the striped orange area (▨) of -1, and the rest of 0; **Right (b)** Exploration sequence $\{w_k\}_{k=0}^{2}$, where $e_1^\top w_k$ and $e_3^\top w_k$ (•), and $e_2^\top w_k$ and $e_4^\top w_k$ (×), respectively.

and final states. Problem (3.30) with $[\alpha_1 \ \alpha_2] = [1 \ 0]$ has a similar structure. Thus, Theorem 4 provides an explicit solution to this state transition problem. Moreover, this viewpoint underlines the relationship to non-commutative maps and flows as indicated in Figure 2.3.

Relation to the Inversion of the Signature of a Path

An important mathematical problem is the inversion of the signature of a path, as presented by Lyons and Xu (2018). The so-called first-level signature coefficients are the increment of the path, and the second level signature collection represents the area confined by the projections of the path on corresponding planes of the given space. This can be continued to the dimension of the space in which the analyzed path is defined. Thus, the third-level signature collection is then given by the enclosed volumes of the path and so forth. For the formal definition of the signature of a path, we refer to Definition 1.2 in Lyons and Xu (2018).

A potential interpretation of the exploration sequence $\{w_k\}_{k=0}^{m-1}$ is that this sequence generates a linear-wise path in $\mathbb{R}^{2n}$ with corner points

$$c_i = \sum_{k=1}^{i} w_{k-1}, \quad c_0 = 0, \quad i = 1, \ldots, m. \tag{3.32}$$

Then the determination of the exploration sequence can be read as a sub-problem of the work from Lyons and Xu (2018) in the following sense: determine a path in $\mathbb{R}^{2n}$ such that the first-level signature coefficients are zero and the second-level signature collection is given by $T(W)$ with the zero vector as an initial point. This formulation is valid for the parameter setting $[\alpha_1 \; \alpha_2] = [1/2 \; 1/2]$ because of Lemma 2. For different α_1, α_2, further investigation into how the projected areas are weighted is necessary. Therefore, our approach presents an algorithmic approach for generating a piecewise linear path for given first- and second-level signature collections.

3.4 Gradient-Generating Functions

This section addresses Step 2) in Section 3.1, i.e., solving the functional equation (3.8) for f, g, and $T_d \in \mathbb{R}^{2n \times 2n}$ with $T(W) = T_d$ in (3.4). First, solutions (T_d, f, g) for the parameter setting $2\alpha_2 - (\alpha_1 + \alpha_2)^2 = 0$ are presented.

Theorem 5. Let $2\alpha_2 - (\alpha_1 + \alpha_2)^2 = 0$ and $T_d \in \mathbb{R}^{2n \times 2n}$ skew-symmetric; then (3.8) is satisfied by the following triples (T_d, f, g), where $a, b \in \mathbb{R}_{>0}$ and $c, \phi \in \mathbb{R}$:

- $T_d = \begin{bmatrix} 0 & -I \\ I & 0 \end{bmatrix}$,

$$g(z) = -f(z) \int f(z)^{-2} dz, \; f : \mathbb{R} \to \mathbb{R}, \tag{3.33}$$

- $T_d = \begin{bmatrix} aQ & -I \\ I & bQ \end{bmatrix}$, $Q = -Q^\top$,

$$f(z) = a^{-1/2} \sin\left(\sqrt{ab}z + \phi\right), \; g(z) = b^{-1/2} \cos\left(\sqrt{ab}z + \phi\right), \tag{3.34}$$

- $T_d = \begin{bmatrix} aQ & -I \\ I & -bQ \end{bmatrix}$, $Q = -Q^\top$,

$$f(z) = \pm a^{-1/2} \cosh\left(\sqrt{ab}z + \phi\right), \; g(z) = \mp b^{-1/2} \sinh\left(\sqrt{ab}z + \phi\right), \tag{3.35}$$

- $T_d = \begin{bmatrix} Q & -I \\ I & 0 \end{bmatrix}$, $Q = -Q^\top$,

$$f(z) = \pm\sqrt{a}, \; g(z) = \mp\frac{z}{\sqrt{a}}, \tag{3.36}$$

- $T_d = \begin{bmatrix} 0 & -I \\ I & Q \end{bmatrix}$, $Q = -Q^\top$,

$$f(z) = \pm\frac{z}{\sqrt{a}}, \; g(z) = \pm\sqrt{a}, \tag{3.37}$$

- $T_d = \begin{bmatrix} 0 & -I - Q \\ I - Q & 0 \end{bmatrix}$, $Q = -Q^\top$,

$$f(z) = \pm \frac{1}{\sqrt{a}} e^{-\frac{a}{2}z}, \ g(z) = \mp \frac{1}{\sqrt{a}} e^{\frac{a}{2}z}, \tag{3.38}$$

- $T_d = \begin{bmatrix} aQ & -I - cQ \\ I - cQ & bQ \end{bmatrix}$, $Q = -Q^\top$,

$$f(z) = \sqrt{\frac{b}{ab - c^2}} \sin\left(\sqrt{ab - c^2} z + \phi\right), \ g(z) = b^{-1/2} \cos\left(\sqrt{ab - c^2} z + \phi\right). \tag{3.39}$$

For each T_d in (3.33) – (3.39) there exists a W, such that $T(W) = T_d$ in (3.4). In (3.39), we require that $a, b > c$. •

The proof of Theorem 5 is given in Appendix B.7.

Remark 3. Every pair f, g in (3.34) – (3.39) satisfies (3.33); thus, these generating functions are also valid for the given T_d in (3.33). The advantages of the specified T_d's are discussed in Chapter 4. •

Remark 4. Consider the indefinite integral in (3.33). Let $F : \mathbb{R}^n \to \mathbb{R}$ be an anti-derivative of $f(z)^{-2}$. Then so is $F + \bar{c}$ for any $\bar{c} \in \mathbb{R}$. Set $g(z) = -f(z)(F(z) + \bar{c})$. The constant $\bar{c}$ is chosen such that $g'(z)f(z) - f'(z)g(z) = -1$. •

Solutions (T_d, f, g) of (3.8) for the parameter setting $2\alpha_2 - (\alpha_1 + \alpha_2)^2 \neq 0$ are presented next.

Theorem 6. Let $T_d \in \mathbb{R}^{2n \times 2n}$ be normal and $(2\alpha_2 - (\alpha_1 + \alpha_2)^2)(T_d + T_d^\top)$ be positive definite; then (3.8) is satisfied by the following triples (T_d, f, g), where $r : \mathbb{R} \to \mathbb{R}_{>0}$, $a \in \mathbb{R} \setminus \{0\}$, $b \in \mathbb{R}_{>0}$, and $\phi \in \mathbb{R}$:

- $T_d = \begin{bmatrix} aI & -I \\ I & aI \end{bmatrix}$, $a(2\alpha_2 - (\alpha_1 + \alpha_2)^2) > 0$,

$$f(z) = \sqrt{r(z)} \sin(\varphi(z)), \ g(z) = \sqrt{r(z)} \cos(\varphi(z)),$$

$$\varphi(z) = \frac{a}{2} \ln(r(z)) + \int \frac{1}{r(z)} dz + \phi, \tag{3.40}$$

- $T_d = \begin{bmatrix} Q & -I \\ I & Q \end{bmatrix}$,

$(2\alpha_2 - (\alpha_1 + \alpha_2)^2)(Q + Q^\top)$ pos. def. and normal,

$$f(z) = b^{-1/2} \sin\left(bz + \phi\right),$$

$$g(z) = b^{-1/2} \cos\left(bz + \phi\right). \tag{3.41}$$

For every T_d in (3.40) – (3.41) there exists a W, such that $T(W) = T_d$ in (3.4). •

The proof of Theorem 6 is given in Appendix B.8.

Remark 5. The list of triples (T_d, f, g) in Theorem 5 is essentially exhaustive, save for some scaled version of the presented cases. A case-by-case study is presented in the proof of Theorem 5 in Appendix B.7. However, the list of triples (T_d, f, g) in Theorem 6 is not exhaustive (cf. Appendix B.8). •

Theorem 5 and Theorem 6, together with Theorem 4, solve (3.8) and (3.7) and thus ensure the existence of an exploration sequence matrix W. Thus, a gradient descent step is approximated by the proposed algorithm (3.1) with transition maps (3.2) as indicated by (1.4).

3.5 Algorithm Design

This section is dedicated to giving interpretations of the algorithm's design parameters and functions as well as some examples and corollaries regarding the exploration sequence. Note that a qualitative and quantitative numerical study of the algorithm's design parameters and functions is part of Chapter 4 and we focus in this section on their theoretical interpretation.

Map parameters α_1, α_2.   The parameters weigh $s_k(J(x_k))$ and $s_k(J(x_k + \sqrt{h}s_k(J(x_k))))$ in (3.2), respectively. In particular, they can be utilized to choose between a single-point ($\alpha_2 = 0$) or a two-point map structure. They are to be normalized as $\alpha_1 + \alpha_2 = 1$ while the ratio of α_1, α_2 can be tuned. Moreover, the choice of α_1, α_2 restricts the choice of T_d to be skew-symmetric for $2\alpha_2 + (\alpha_1 + \alpha_2)^2 = 0$ and otherwise normal with $(2\alpha_2 - (\alpha_1 + \alpha_2)^2)(T_d^\top + T_d)$ positive definite (cf. Theorem 5 and Theorem 6).

Gradient-generating functions f, g.   Various choices are presented in Theorem 5 and Theorem 6, depending on α_1, α_2. It is worth mentioning that it can be advantageous if f, g are chosen as sinusoidal functions because of their bounded appearances. Note that if f, g have high function values, or if f, g scale arbitrarily large with $J(x_k)$, the algorithm performs large steps that may cause instabilities and divergence. Further, if J, f, g vanish at a minimum x^*, asymptotic convergence to x^* (instead of practical asymptotic convergence) can be achieved.

Exploration sequence matrix W and T_d.   The exploration sequence matrix W depends on the choice of T_d, specifically on the eigenvalues of T_d (see Example 2 – Example 5; p. 31ff). A step-by-step construction of W based on the algorithm parameters is presented in Appendix D.1. As explained in this construction, the singular values of W can be chosen (see also Corollary 2; p. 34), hence, this degree of freedom can be used in the algorithm design and tuning (see Example 6 and Example 7; p. 34ff). As discussed below, smaller singular values of W lead to a larger exploration sequence length m. However, there exists a set of optimal singular values of W w.r.t. the choice of T_d in the sense of minimal number of steps m, which is $m = \mathrm{rk}(T_d) + 1$. Therewith, the choice of T_d influences the lower bound on m (see Corollary 3; p. 34).

Step size h. The approximated gradient is scaled with the step size h; thus , h influences the speed of convergence as well as the area of exploration around x_k. As stated in Theorem 2, there exists an upper bound on h such that convergence is ensured in theory. In practical applications, larger step sizes can be chosen. We will discuss an adaptive and variable step size in Section 4.4.

In the sequel of this section we discuss the design of the exploration sequence matrix W in more detail. We provide several examples to investigate the influence on the exploration sequence of α_1, α_2, and T_d, as well as the singular values of W denoted by $\sigma_\ell \in \mathbb{R}_{>0}$, $\ell = 1, \ldots, \mathrm{rk}(T_d)$. The first two examples apply Theorem 4 and illustrate the dependency on T_d.

Example 2. Let $n = 1$, $[\alpha_1\ \alpha_2] = [1/2\ 1/2]$, and

$$T_d = \begin{bmatrix} 0 & -1 \\ 1 & 0 \end{bmatrix}, \tag{3.42}$$

i.e., $\mathrm{rk}(T_d) = 2$. Then following the constructive proof of Theorem 4 in Appendix B.5 or the step-by-step construction of W in Appendix D yields

$$W = \frac{1}{\sqrt[4]{3}} \begin{bmatrix} \sqrt{2-\sqrt{3}} & -\sqrt{2+\sqrt{3}} & \sqrt{2} \\ \sqrt{2+\sqrt{3}} & -\sqrt{2-\sqrt{3}} & -\sqrt{2} \end{bmatrix} \in \mathbb{R}^{2\times 3}. \tag{3.43}$$

Thus, $m = \mathrm{rk}(T_d) + 1 = 3$. Moreover, it is simple to check that (3.21) is satisfied.   •

Example 3. Let $n = 2$, $[\alpha_1\ \alpha_2] = [1/2\ 1/2]$, and

$$T_d = \begin{bmatrix} 0 & -1 & -1 & 0 \\ 1 & 0 & 0 & -1 \\ 1 & 0 & 0 & -1 \\ 0 & 1 & 1 & 0 \end{bmatrix}, \tag{3.44}$$

i.e., $\mathrm{rk}(T_d) = 2$. Then as in Example 2, the construction of W leads to

$$W = \frac{1}{\sqrt[4]{3}} \begin{bmatrix} -\sqrt{2+\sqrt{3}} & \sqrt{2-\sqrt{3}} & \sqrt{2} \\ \sqrt{2-\sqrt{3}} & -\sqrt{2+\sqrt{3}} & \sqrt{2} \\ \sqrt{2-\sqrt{3}} & -\sqrt{2+\sqrt{3}} & \sqrt{2} \\ \sqrt{2+\sqrt{3}} & -\sqrt{2-\sqrt{3}} & -\sqrt{2} \end{bmatrix} \in \mathbb{R}^{4\times 3}. \tag{3.45}$$

Thus, $m = \mathrm{rk}(T_d) + 1 = 3$. Comparing (3.45) with (3.12) reveals a reduction of m by more than factor two. Therefore, choosing T_d in a proper way can reduce the sequence length m and a gradient approximation in less steps.   •

m	ω_1	ω_2	ω_3	ω_4	ω_5
3	0.2887	-	-	-	-
4	0.5	0	-	-	-
5	0.6882	0.1625	-	-	-
6	0.8660	0.2887	0	-	-
7	1.0382	0.3987	0.1141	-	-
8	1.2071	0.5	0.2071	0	-
9	1.3737	0.5958	0.2886	0.0881	-
10	1.5388	0.6882	0.3633	0.1625	0
11	1.7028	0.7780	0.4333	0.2283	0.0719

Table 3.1. Eigenvalues of $\bar{P}(m) \in \mathbb{R}^{(m-1)\times(m-1)}$ in (3.26)

Example 2 and Example 3 highlight the minimum sequence length m w.r.t T_d as stated in Corollary 1. The next two examples focus on how Conjecture 1 plays a role in Theorem 4. Before we present the examples, it is necessary to state the conditions which must be satisfied by the eigenvalues of T_d and $\bar{P}(m)$ in (3.26). Let the eigenvalues of T_d be $\gamma_\ell \pm \delta_\ell i$, $\gamma_\ell \in \mathbb{R}$, $\delta_\ell \in \mathbb{R}_{\geq 0}$ for $\ell = 1,\ldots,\lceil n/2 \rceil$ with $\gamma_\ell = 0$ and $\delta_\ell = 0$ for $\ell > \lceil \mathrm{rk}(T_d)/2 \rceil$; the eigenvalues of $\bar{P}(m)$ be $1/2(\alpha_1 + \alpha_2^2) - \alpha_2 \pm \omega_k i$, $\omega_k \in \mathbb{R}_{\geq 0}$ for $k = 1,\ldots,\lfloor m/2 \rfloor$; and the eigenvalues of the principal submatrix $\bar{P}(m)_{1:\mathrm{rk}(T_d)}$ of $\bar{P}(m)$ be $1/2(\alpha_1 + \alpha_2)^2 - \alpha_2 \pm \hat{\omega}_k i$, $\hat{\omega}_k \in \mathbb{R}_{\geq 0}$ for $k = 1,\ldots,\lfloor \mathrm{rk}(T_d)/2 \rfloor$. In light of Theorem 4, the interlacing eigenvalue property (cf. Lemma 5 in Appendix A.2) of $\bar{P}(m)$, such as

$$\omega_k \geq \hat{\omega}_k \geq \omega_{\lceil(m-1)/2\rceil - \lceil\mathrm{rk}(T_d)/2\rceil + k}, \qquad k = 1,\ldots,\lceil \mathrm{rk}(T_d)/2 \rceil \tag{3.46}$$

must be satisfied if $\gamma_k \neq 0$ (case where $2\alpha_2 - (\alpha_1 + \alpha_2)^2 \neq 0$, i.e., T_d's in Theorem 6) by

$$\hat{\omega}_k = \frac{\delta_k}{\gamma_k}\left(\frac{1}{2}(\alpha_1 + \alpha_2)^2 - \alpha_2\right), \qquad k = 1,\ldots,\lceil \mathrm{rk}(T_d)/2 \rceil. \tag{3.47}$$

For more details and the derivation of the condition (3.47), we refer to the proof of Theorem 4 in Appendix B.5. In this view and in combination with Conjecture 1, m must be chosen large enough such that for every $\hat{\omega}_k$ in (3.47) the interlacing eigenvalue property (3.46) holds. The imaginary part ω_k of $\bar{P}(m)$'s eigenvalues for $m = 3,\ldots,11$ are stated in Table 3.1. Note that the interlacing eigenvalue property of $\bar{P}(m)$, i.e., (3.25) is apparent by comparing the rows in Table 3.1. Specifically, the largest eigenvalue ω_1 is increasing, while the smallest eigenvalue is either zero or decreasing (for m even) in regards of m.

Example 4. Let $n = 2$, $[\alpha_1\ \alpha_2] = [1\ 0]$, and

$$T_d = \begin{bmatrix} -I & -I \\ I & -I \end{bmatrix}, \tag{3.48}$$

i.e., $\mathrm{rk}(T_d) = 4$. Then following the constructive proof of Theorem 4 in Appendix B.5 or the step-by-step construction of W in Appendix D yields

$$W = \begin{bmatrix} -\sqrt{2} & 0 & \sqrt{2} & 0 & -\sqrt{2} & 0 & \sqrt{2} & 0 \\ -0.5 & \sqrt{2} & -0.5 & 0 & 0.5 & -\sqrt{2} & 0.5 & 0 \\ 0.5 & 0 & -0.5 & \sqrt{2} & -0.5 & 0 & 0.5 & -\sqrt{2} \\ 0 & \sqrt{2} & 0 & -\sqrt{2} & 0 & \sqrt{2} & 0 & -\sqrt{2} \end{bmatrix} \in \mathbb{R}^{4 \times 8}, \tag{3.49}$$

i.e., $m \neq \mathrm{rk}(T_d) + 1 = 5$. Reviewing the eigenvalues of T_d, given by the double eigenvalue pairs $-1 \pm i$, leads to $\hat{\omega}_1 = \hat{\omega}_2 = 0.5$. Therefore, based on (3.46) and (3.47), the inequalities

$$\omega_1 \geq 0.5 \geq \omega_{\lceil (m-1)/2 \rceil - 1} \quad \text{and} \quad \omega_2 \geq 0.5 \geq \omega_{\lceil (m-1)/2 \rceil} \tag{3.50}$$

must be satisfied. Obviously, according to Table 3.1, this is fulfilled for $m \geq 8$.   •

Example 5. Let $n = 2$, $[\alpha_1\ \alpha_2] = [1\ 0]$, and

$$T_d = \begin{bmatrix} -\sqrt{5 + 2\sqrt{5}} & 0 & -1 & 0 \\ 0 & -\sqrt{5 - 2\sqrt{5}} & 0 & -1 \\ 1 & 0 & -\sqrt{5 + 2\sqrt{5}} & 0 \\ 0 & 1 & 0 & -\sqrt{5 - 2\sqrt{5}} \end{bmatrix}, \tag{3.51}$$

i.e., $\mathrm{rk}(T_d) = 4$. Then as in Example 4, the construction of W leads to

$$W = \begin{bmatrix} 0.3374 & 0.1289 & -0.5459 & 0.7545 & -0.6748 \\ -0.2653 & -1.5528 & -0.6944 & 1.1236 & 1.3889 \\ -0.6837 & 0.7514 & -0.5322 & 0.1096 & 0.3548 \\ 1.5465 & 0.2256 & -1.4071 & -1.0953 & 0.7302 \end{bmatrix} \in \mathbb{R}^{4 \times 5}. \tag{3.52}$$

Reviewing the eigenvalues of T_d, given by $-3.0777 \pm i$ and $-0.7265 \pm i$, leads to $\hat{\omega}_1 = 0.6882$ and $\hat{\omega}_2 = 0.1625$. With the same lines of argument as in Example 4, (3.46) is satisfied with $m \geq 5$. Hence, m depends on T_d, specifically on its rank and eigenvalues. We elaborate on the choice of T_d in Section 3.4.   •

As stated in Theorem 4, the satisfaction of Conjecture 1 is necessary for $2\alpha_2 - (\alpha_1 + \alpha_2)^2 \neq 0$. However, this property can be utilized for design purpose for any α_1, α_2 setting (see Corollary 2). Based on the proof of Theorem 4 in Appendix B.5, the singular values $\sigma_\ell \in \mathbb{R}_{>0}$, $\ell = 1, \ldots, \mathrm{rk}(T_d)$ of W must satisfy (3.47) and the additional conditions

$$\sigma_{2k-1} \sigma_{2k} \hat{\omega}_k = \delta_k \tag{3.53}$$

for $k = 1, \ldots, \lceil \mathrm{rk}(T_d)/2 \rceil$. Either $\sigma_{2k-1}, \sigma_{2k}$ are calculated based on $\hat{\omega}_k$ and δ_k such that the minimum sequence length m for given α_1, α_2 and T_d is achieved (see Corollary 3), or $\sigma_{2k-1}, \sigma_{2k}$ are chosen arbitrarily for design purposes. How this influences W is discussed in the next examples and from a numerical perspective in Chapter 4.

Corollary 2. In the case $2\alpha_2 - (\alpha_1 + \alpha_2)^2 = 0$, the singular values $\sigma_{2\ell-1}, \sigma_{2\ell}, \ell = 1, \ldots, \mathrm{rk}(T_d)$ of W can be chosen arbitrarily. Otherwise, the singular values of W must satisfy $\sigma_{2\ell-1} = \sigma_{2\ell}$, $\ell = 1, \ldots, \lceil \mathrm{rk}(T_d)/2 \rceil$ and

$$T_d = \begin{bmatrix} \mathrm{diag}([\gamma_1 \cdots \gamma_n]) & -I \\ I & \mathrm{diag}([\gamma_1 \cdots \gamma_n]) \end{bmatrix}, \tag{3.54}$$

with

$$\gamma_\ell = \left(\alpha_2 - \frac{1}{2}(\alpha_2 + \alpha_1)^2\right)\sigma_{2\ell-1}^2, \tag{3.55}$$

must hold. ●

Proof. Following directly from the proof of Theorem 4 in Appendix B.5, specifically, T_d in (3.54) satisfies (3.41) and (3.55) corresponds to (B.26). □

Corollary 3. The minimal number of steps necessary to approximate a gradient step according to (3.1) and (3.2) and using the matrices T_d in Theorem 5 and Theorem 6 is $m = n + 1$. ●

Proof. Since the first n rows of each T_d in Theorem 5 and Theorem 6 are linearly independent, we know $\min\{\mathrm{rk}(T_d)\} \geq n$. This implies with Theorem 4 that $m \geq n + 1$. □

Remark 6. A gradient step is approximated in $m = n + 1$ steps for $[\alpha_1\ \alpha_2] = [1/2\ 1/2]$ with T_d as specified in (3.34) where Q has elements $\{q_{ij}\}_{i,j=1}^{2n}$ such that

$$q_{ij} = \begin{cases} 1 & \text{if } i + j = 2n + 1, \ i > j \\ -1 & \text{if } i + j = 2n + 1, \ i < j \\ 0 & \text{else} \end{cases} \tag{3.56}$$

hold, while the singular values $\sigma_{2\ell-1}, \sigma_{2\ell}$ of W satisfy (3.53) for $\ell = 1, \ldots, \lceil n/2 \rceil$. ●

Remark 7. In Example 2, the singular values $\sigma_1 = \sigma_2 = \sqrt{2}\sqrt[4]{3}$ are calculated such that with (3.53), $\hat{\omega}_1 = (2\sqrt{3})^{-1}$. Therefore, the minimum sequence length $m = 3$ w.r.t. T_d and α_1, α_2 is achieved (see Table 3.1). ●

Example 6. Let n, α_1, α_2, and T_d be as in Example 2. Moreover, we choose the singular values $\sigma_1 = \sigma_2 = 1$, i.e., with (3.53) the imaginary parts of the eigenvalues of the principal submatrix $\bar{P}(m)_{1:2}$ must be $\hat{\omega}_1 = 1$. Therefore, with Table 3.1 and (3.46), the sequence length is $m = 7$. Then following the constructive proof of Theorem 4 in Appendix B.5 or the step-by-step construction of W in Appendix D.1 yields

$$W = \begin{bmatrix} -0.5622 & -0.0185 & 0.4568 & 0.4281 & 0.2306 & -0.0489 & -0.4859 \\ -0.2181 & -0.6287 & -0.3803 & 0.1873 & 0.4137 & 0.3848 & 0.2414 \end{bmatrix} \in \mathbb{R}^{2\times 7}. \tag{3.57}$$

●

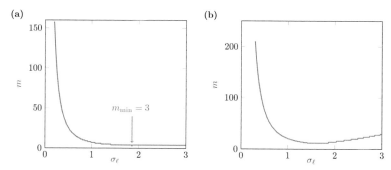

Figure 3.3. Dependency of the singular values σ_ℓ, $\ell = 1, \ldots, \mathrm{rk}(T_d)$ of the exploration sequence matrix $W \in \mathbb{R}^{2n \times m}$ on the sequence length m, where $\sigma_i = \sigma_j$ for $i, j = 1, \ldots, \mathrm{rk}(T_d)$. **Left (a)** Setup: $n = 1$, $\alpha_1 = \alpha_2 = 0.5$, and $T_d = [0 \;-1;\; 1\; 0] \in \mathbb{R}^{2 \times 2}$; **Right (b)** Setup: $n = 3$, $\alpha_1 = \alpha_2 = 0.5$, and $T_d = [0 \;-I;\; I\; 0] \in \mathbb{R}^{6 \times 6}$. Note that $m_{\min} = 7$ for the given setting is achieved for $\sigma_1 = \sigma_2 = 2.96$, $\sigma_3 = \sigma_4 = 1.58$, and $\sigma_5 = \sigma_6 = 0.98$, while for all $\sigma_\ell = 1.7$ the smallest exploration sequence length is $m = 12$.

If the singular values of W are chosen larger than the ones corresponding to the minimum sequence length m based on the given T_d and α_1, α_2, the sequence length increases. The dependency of the singular values of W on the sequence length m for the special case of identical singular values is depicted in Figure 3.3.

Moreover, in addition to influencing the sequence length m, the singular values of W define the amplitude ratio of each sequence $\{e_i^\top w_k\}_{k=0}^{m-1}$ for $i = 1, \ldots, 2n$ as the next corollary and example present:

Corollary 4. Let the eigenvectors of T_d be given as $a_\ell \pm b_\ell i$, $a_\ell, b_\ell \in \mathbb{R}^{2n}$, $\ell = 1, \ldots, n$ and the singular values of W as $\sigma_1 \geq \sigma_2 \geq \ldots \geq \sigma_{\mathrm{rk}(T_d)} \geq 0 = \ldots = 0$ with $\sigma_k \in \mathbb{R}_{\geq 0}$ for $k = 1, \ldots, 2n$. Then it holds that

$$\|e_i^\top W\|_2 = |e_i^\top \sum_{j=1}^{\lceil \mathrm{rk}(T_d)/2 \rceil} (a_j \sigma_{2j-1} + b_j \sigma_{2j})|. \tag{3.58}$$

•

Corollary 4 follows directly from the proof of Theorem 4 in Appendix B.5.

Remark 8. The evaluation maps $s_\ell(J(x))$ defined in (3.2) can be expressed in terms of W as

$$s_\ell(J(x)) = Y(f(J(x)), g(J(x))) W_{k \bmod(m)+1}. \tag{3.59}$$

Then Corollary 4 implies that the scaling in each dimension $j = 1, \ldots, n$ depends on $\|e_j^\top W\|_2$ and $\|e_{j+n}^\top W\|_2$. This is illustrated in Example 7 and in Section 4.2 from a numerical perspective. •

Example 7. Let n, α_1, α_2, and T_d be as in Example 3. Moreover, we choose the singular values $\sigma_1 = 2$, $\sigma_2 = 1$, i.e., with (3.53) the imaginary parts of the eigenvalues of the principal submatrix $\check{P}(m)_{1:2}$ has to be $\hat{\omega}_1 = 1$ (note that $\delta_1 = 2$). Therefore, with Table 3.1 and (3.46), the sequence length is $m = 7$. Then following the constructive proof of Theorem 4 in Appendix B.5 or the step-by-step construction of W in Appendix D.1 yields

$$W = \begin{bmatrix} 0.1542 & 0.4446 & 0.2689 & \text{-}0.1324 & \text{-}0.2925 & \text{-}0.2721 & \text{-}0.1707 \\ \text{-}0.7951 & \text{-}0.0262 & 0.6460 & 0.6055 & 0.3262 & \text{-}0.0691 & \text{-}0.6872 \\ \text{-}0.7951 & \text{-}0.0262 & 0.6460 & 0.6055 & 0.3262 & \text{-}0.0691 & \text{-}0.6872 \\ \text{-}0.1542 & \text{-}0.4446 & \text{-}0.2689 & 0.1324 & 0.2925 & 0.2721 & 0.1707 \end{bmatrix} \in \mathbb{R}^{2 \times 7}. \quad (3.60)$$

Moreover, the eigenvectors $a_\ell \pm b_\ell i$, $a_\ell, b_\ell \in \mathbb{R}^{2n}$, $\ell = 1, 2$ of T_d are given by

$$\begin{bmatrix} a_1 & b_1 & a_2 & b_2 \end{bmatrix} = \begin{bmatrix} 0 & 0.5 & 0 & 0 \\ 0.5 & 0 & -\sqrt{2}/2 & 0 \\ 0.5 & 0 & \sqrt{2}/2 & 0 \\ 0 & -0.5 & 0 & 0 \end{bmatrix}. \quad (3.61)$$

Thus, $\|e_1^\top W\|_2 = \|e_4^\top W\|_2 = 0.5\sigma_2 = 0.5$ and $\|e_2^\top W\|_2 = \|e_3^\top W\|_2 = 0.5\sigma_1 = 1$. Therefore, the singular values can be also utilized for tuning the ratio of each sequence $\{e_i^\top w_k\}_{k=0}^{m-1}$ for $i = 1, \ldots, 2n$. ●

3.6 Conclusion and Outlook

In this chapter we performed a theoretical analysis of the presented algorithm class. First, a parameterized version of the non-commutative maps was introduced, that involved gradient-generating functions and periodic exploration sequences. The main part of this chapter was dedicated to the derivation and design of these ingredients. We constructed the exploration sequence based on the solution of a system of quadratic equations, which we underpinned with a geometric interpretation. Furthermore, we drew several relations of the exploration sequence construction to contiguous research areas, specifically to non-holonomic systems and the inversion of the signature of a path. For the determination of the gradient-generating functions a set of functional equations was solved. We presented various pairs of gradient-generating functions that will be analyzed from a numerical perspective and their influence on the algorithm's performance in the next chapter. Further, we proved semi-global practical asymptotic and semi-global asymptotic convergence—if a decreasing step size is applied—of the algorithm class. Eventually, we presented first design parameter choices for the presented algorithms, which we will discuss from a numerical perspective in the next chapter.

Outlook. One clear further research direction is the investigation and proof of Conjecture 1. We validated it numerically for matrices in $\mathbb{R}^{m \times m}$ with $m \leq 10000$, which is sufficient for optimization problems for $n \leq 2500$ (known coordinate-wise exploration sequence (3.11) with $m = 4n$). One starting point to tackle the proof might be to consider the characteristic polynomial $p(m, \lambda)$ with eigenvalues $\lambda \in \mathbb{C}$ of (3.22). Then it remains open to show that the polynomials $p^2(m, \lambda)$ and $p^2(m + 1, \lambda)$ have interlacing roots for any m (the squared polynomials must be considered because $C(m)$ in (3.22) has only imaginary eigenvalues). The characteristic polynomial of (3.22) is provided in Appendix A.3.2.

Investigating the relation of the exploration sequence construction to nonholonomic systems (as outlined in Section 3.3) raises further research possibilities. Firstly, a generalization of the non-commutative map approximation scheme to more than two gradient-generating functions such that iterated Lie-brackets are approximated (cf. Sussmann and Liu (1991)). Secondly, studying the exploration sequence and its construction in the limit case, i.e., if the singular values of W tend to zero. We observed that for small singular values the exploration sequence approximates sinusoidal functions, which we know are optimal in the case of the continuous-time non-nonholonomic integrator system (cf. R. Brockett (2014) and (3.31)).

A further interesting direction could be a generalization of the constructive and algorithmic approach to determine the exploration sequence, i.e., constructing a path with given first- and second-level signatures. First, it might be a potential performance tuning of the presented algorithm class to force the exploration sequence to generate a path where the third- or even higher-level signatures are zero and thereby the remainder term is reduced. Thus, the gradient approximation becomes more accurate. Second, a general and algorithmic characterization of solutions to this inverse path problem (as outlined in Section 3.3) with given signature values is not known, and in our opinion it is an interesting mathematical research question by its own (cf. Lyons and Xu (2018)).

Eventually, applying and tailoring our gradient approximation procedure to gradient descent algorithms with momentum like the Heavy Ball method (Polyak (1987)), Nesterov's method (Nesterov (2003)), or the triple momentum method (Van Scoy, Freeman, and Lynch (2017)) with faster convergence rates could be of great value.

4

Numerical Results and Experiments

In this chapter we focus on the implementation and numerical results of the presented algorithm class. At first, the algorithm's implementation, as well as two versions of the algorithm setup that are utilized as prime examples for the numerical studies in the sequel, are specified (Section 4.1). This is followed by a numerical analysis of our algorithm class. In the first part of this analysis the algorithms are applied on the two-dimensional spherical test function $(J(x) = x^\top x)$. Therein, a qualitative and quantitative parameter study (Section 4.2 and Section 4.3) of the algorithm's design parameters and functions (see Section 3.5) is performed. Based on those results, various performance tuning rules (Section 4.4) are discussed and compared within the algorithm class. In the second part of the algorithm's numerical analysis, several simulation experiments are carried out to show its practicality on multi-variable, non-convex, discontinuous, and ill-conditioned objective functions. In essence, comparisons with four other derivative-free optimization algorithms applied on challenging benchmarking problems are presented (Section 4.5). Eventually, we introduce two potential applications of our algorithm class in the field of reinforcement learning (Section 4.6).

Additionally, we entitle the algorithm class as McOpt algorithms (see Figure 4.1)—standing for **M**agic **C**ube **Opt**imization algorithms[1]. This name is given because the magic cube is based on non-commutative operations like the proposed class of algorithms, as outlined in Section 1.1.

Figure 4.1. The presented algorithm class in this thesis is entitled McOpt algorithms—**M**agic **C**ube **Opt**imization algorithms (image modified under the *GNU General Public License* (2020)).

---

[1]A unilateral relation to the music industry, where "Mc" (Microphone Controller) is a common naming term for DJs, can be drawn. The objective of DJs is to maximize the temper of their crowd, where only current uncertain measurements of the mood are available to the DJs. Based on that, the next song must be chosen with the goal to increase or keep the mood level. In this view, DJs must solve a derivative-free optimization problem with a time-dependent objective function to be maximized

4.1 Algorithm Implementation and Setup

The design parameters and functions of McOpt must be chosen in the following order:

1. Map parameters $\alpha_1, \alpha_2 \in \mathbb{R}$ with $\alpha_1 + \alpha_2 \neq 0$

2. Exploration sequence matrix $W \in \mathbb{R}^{2n \times m}$, in particular matrix $T_d \in \mathbb{R}^{2n \times 2n}$ based on Theorem 5 and Theorem 6, as well as singular values $\sigma = \{\sigma_1, \sigma_2, \ldots, \sigma_{\mathrm{rk}(T_d)}\}$ of W as described in Section 3.5. For the construction of W, see Appendix D.1.

3. Gradient-generating functions $f, g : \mathbb{R} \to \mathbb{R}$, based on Theorem 5, Theorem 6, and the choice of T_d.

4. Step size $h \in \mathbb{R}_{>0}$.

For the numerical studies in the sequel, we introduce the so called *algorithm characteristics* specified by the map parameters α_1, α_2. In particular, we define the

[C1] one-point characteristic $[\alpha_1 \ \alpha_2] = [1 \ 0]$ and

[C2] two-point characteristic $[\alpha_1 \ \alpha_2] = [1/2 \ 1/2]$.

As previously mentioned, [C1] yields one function evaluation and [C2] two function evaluations per iteration. Note that in this parameterized setting no other single-point algorithm exists except scaled versions by α_1. Therewith, McOpt with algorithm characteristic [C1] is given in terms of the exploration matrix W by

$$x_{k+1} = x_k + \sqrt{h} Y\big(f(J(x_k)), g(J(x_k))\big) W e_{k \bmod(m)+1}, \quad k \geq 0 \tag{4.1}$$

and McOpt with algorithm characteristic [C2] by

$$x_{k+1} = x_k + \frac{1}{2}\sqrt{h}\Big(Y\big(f(J(x_k)), g(J(x_k))\big) + Y\big(f(J(\hat{x}_k)), g(J(\hat{x}_k))\big)\Big) W e_{k \bmod(m)+1}, \tag{4.2a}$$

$$\hat{x}_k = x_k + \sqrt{h} Y\big(f(J(x_k)), g(J(x_k))\big) W e_{k \bmod(m)+1}, \quad k \geq 0 \tag{4.2b}$$

with

$$Y\big(f(z), g(z)\big) = [f(z)I \ \ g(z)I] \in \mathbb{R}^{n \times 2n}. \tag{4.3}$$

In summary, for general α_1, α_2 the algorithm is implemented as presented in Algorithm 1 (p. 41).

4.2 Design Parameters and Functions: Qualitative Study

In this section we discuss the influence of the McOpt algorithm's design parameters and functions for the two setups (4.1) and (4.2), in particular, the gradient-generating functions

---

Algorithm 1 McOpt

1: **Input:** $x_0, h, \alpha_1, \alpha_2, f(\cdot), g(\cdot), T_d, \sigma_i$ $(i = 1, \ldots, \text{rk}(T_d))$, stop criterion
2: Calculate W and m as described in Appendix D.1
3: $k = 0$
4: **while** stop criterion is not fulfilled **do**
5: $\quad \ell = k \bmod (m) + 1$
6: $\quad e_\ell = [0_\ell, 1, 0_{n-1-\ell}]^\top$
7: $\quad Y_k = [f(J(x_k))I \; g(J(x_k))I]$
8: $\quad \hat{x}_k = x_k + \sqrt{h} Y_k W e_\ell$
9: $\quad \hat{Y}_k = [f(J(\hat{x}_k))I \; g(J(\hat{x}_k))I]$
10: $\quad x_{k+1} = x_k + \sqrt{h}(\alpha_1 Y_k + \alpha_2 \hat{Y}_k) W e_\ell$
11: $\quad k \leftarrow k + 1$
12: **end while**
13: **return** $[x_0, x_1, \ldots]$

---

f, g, matrix T_d, the singular values $\{\sigma_i\}_{i=1}^{\text{rk}(T_d)}$ of W, and the step size h. Our conclusions are drawn from a qualitative numerical perspective. Hence, we discuss the influence of the design parameters on the algorithm's convergence and exploration behavior. The term "exploration behavior" describes the points that are visited by the sequence $\{x_k\}$ while approaching the neighborhood of a minimum x^*. A more extensive exploration behavior of the algorithm can be of advantage, especially for challenging objective functions, e.g. overcoming local minima. Moreover, we will observe that the generating areas by $T(W)$, as discussed in Section 3.3 and depicted in Figure 3.1, relate to the algorithm's exploration behavior.

In the following studies, various algorithm setups are applied to the two-dimensional spherical test function

$$J(x) = x^\top x + 1, \quad x \in \mathbb{R}^2 \tag{4.4}$$

with global minimum $x^* = [0\,0]^\top$ and $J(x^*) = 1$ as depicted in Figure 4.2. Note that we use a rather simple test function for the sake of comparability of the various algorithm design parameters and functions. In a later stage (Section 4.5), we apply McOpt to more challenging benchmarking test functions, including multi-variable, ill-conditioned, non-convex, and discontinuous objective functions.

In the sequel of this chapter, multiple *Algorithm Setups* referenced in different studies, are introduced to define the algorithm's design parameters and functions. The first considered setup is as follows:

Algorithm Setup 1. McOpt with characteristic **[C2]** corresponding to (3.33) and (3.34) with

- Generating functions: $f(z) = a^{-1/2} \sin(\sqrt{ab}z + \phi)$, $g(z) = b^{-1/2} \cos(\sqrt{ab}z + \phi)$

- $T_d = \begin{bmatrix} aQ & -I \\ I & bQ \end{bmatrix}$ with $Q = Q_1 = 0 \in \mathbb{R}^{2 \times 2}$ or $Q = Q_2 = \begin{bmatrix} 0 & -1 \\ 1 & 0 \end{bmatrix}$

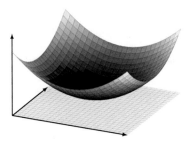

Figure 4.2. Spherical test function $J(x) = x^\top x + 1$ with $x^* = [0\ 0]^\top$ and $J(x^*) = 1$.

where $a, b \in \mathbb{R}_{>0}$ and $\phi \in \mathbb{R}$. The initial value is set to $x_0 = [1\ 1]^\top$, and the step size to $h = 0.05$.   •

The default setup of parameters is $a = b = 1$, $\phi = 0$, and $Q = Q_1$; they are chosen if not otherwise specified. In the following we discuss every design parameter separately and highlight the observed numerical results of McOpt.

Singular Values of W

As elaborated in Section 3.5 the choice of singular values of W influence the sequence length m; smaller values lead to a larger m, and there exists a minimal m w.r.t. T_d. In our first study the default design parameters with $Q = Q_2$ in Algorithm Setup 1 (p. 41) are selected with three different singular value settings. Note that in this setup $\mathrm{rk}(T_d) = 2$, i.e., two singular values $\sigma = \{\sigma_1, \sigma_2\}$ must be chosen. Moreover, W and m are determined based on the construction procedure of the exploration matrix in Appendix D.1. The choices are (a) $\sigma = \{2\sqrt[4]{3}, 2\sqrt[4]{3}\}$, which yields the minimal sequence length $m = 3$; (b) $\sigma = \{0.7, 0.7\}$, to investigate the behavior for small singular values, which leads to $m = 26$; and (c) $\sigma = \{2.5, 0.5\}$, to study how each singular value affects the exploration, reveals $m = 11$. A visualization of the evolution of x_k, $J(x_k)$, and the area generating effects as described in Section 3.3 are depicted in Figure 4.3.

Obviously, the convergence speed depends on the exploration sequence length m, i.e., the number of steps to approximate the gradient of J as visible in the center plots of Figure 4.3. For smaller singular values, the shape of the generated areas tends to be more "round" (cf. Figure 4.3 (a) and Figure 4.3 (b)), and the evolution of x_k and $J(x_k)$ generates a "smoother" trajectory. If the singular values chosen are not identical as in (c), the shapes of the generated areas by $T(W)$ are given in elongated elliptic forms (see Figure 4.3 (c)). This implies that each exploration sequence $\{e_i^\top w_k\}_{k=0}^{m-1}$ for $i = 1, \ldots, 4$ is weighted differently as stated in

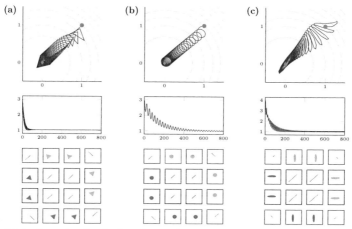

Figure 4.3. An illustration showing how the singular values of W influence the McOpt algorithm's behavior. McOpt with parameters, as in Algorithm Setup 1 (p. 41) with $Q = Q_2$, is applied to (4.4). The vertical alignment of the figure is as follows: **(Top)** Visualization of $x_k \in \mathbb{R}^2$; **(Center)** Objective value $J(x_k)$ over k; **(Bottom)** Area plot as described in Section 3.3. The singular values of W are chosen as follows: **Left (a)** $\sigma = \{2\sqrt[4]{3}, 2\sqrt[4]{3}\}$ yields $m = 3$; **Middle (b)** $\sigma = \{0.7, 0.7\}$ yields $m = 26$; **Right (c)** $\sigma = \{2.5, 0.5\}$ yields $m = 11$.

Corollary 4. Specifically, in this case

$$\|e_1^\top W\|_2 = \frac{\sigma_2}{\sigma_1}\|e_2^\top W\|_2 \quad \text{and} \quad \|e_3^\top W\|_2 = \frac{\sigma_1}{\sigma_2}\|e_4^\top W\|_2, \tag{4.5}$$

which yields to the elongated elliptic exploration of the sequence $\{x_k\}$.

Matrix T_d

The choice of T_d, particularly its eigenvalues and eigenvectors, influences W and the sequence length m, as discussed in Section 3.3. To investigate various settings of T_d, we choose the default design parameters in Algorithm Setup 1 (p. 41) with singular values $\sigma = \{1, 1, 1, 1\}$ of W, and we set T_d to (a) $Q = Q_1$; (b) $Q = 3Q_2$; and (c) $Q = 0.2Q_2$. The algorithm's behavior, i.e., the evolution of x_k, $J(x_k)$, and the generated areas are visualized in Figure 4.4. The sequence length m is given by $m = 14$ in (a), $m = 26$ in (b), and $m = 12$ in (c). This corresponds to the scale of the largest eigenvalue of T_d (see Example 2 – Example 5); for the given settings, T_d has its largest eigenvalue for (b). Additionally, the elements of T_d define the surface area of the generated areas by $T(W)$ as depicted in the bottom plots of Figure 4.4. We observe that larger matrix elements correlate to a more extensive exploration

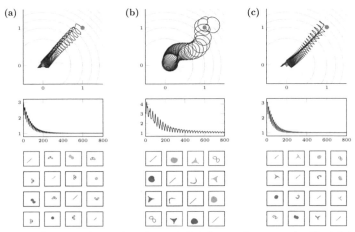

Figure 4.4. An illustration showing how the matrix T_d influences the McOpt algorithm's behavior. McOpt with parameters as in Algorithm Setup 1 (p. 41) with singular values $\sigma = \{1, 1, 1, 1\}$ of W is applied to (4.4). The vertical alignment of the figure is as follows: **(Top)** Visualization of $x_k \in \mathbb{R}^2$; **(Center)** Objective value $J(x_k)$ over k; **(Bottom)** Area plot as described in Section 3.3. The matrix T_d as in Algorithm Setup 1 (p. 41) is chosen as follows: **Left (a)** T_d with $Q = Q_1$ yields $m = 14$; **Middle (b)** T_d with $Q = 3Q_2$ yields $m = 26$; **Right (c)** T_d with $Q = 0.2Q_2$ yields $m = 12$.

behavior, i.e., the sequence $\{x_k\}$ covers a larger area of exploration while converging to the neighborhood of x^*. As previously mentioned, this can be of advantage for more challenging objective functions, for example with local minima. However, tuning T_d is not that intuitive, and the behavior is difficult to predict. Therefore, our recommended choice is to choose T_d w.r.t Theorem 5 and Theorem 6 such that the lowest possible rank is given and a small exploration sequence length m can be gained.

Parameters of Sinusoidal Gradient-Generating Functions f, g

In this paragraph we analyze the influence of the parameters a, b, ϕ in Algorithm Setup 1 (p. 41). Clearly, if $Q = Q_1$ is chosen, a, b solely scale the gradient-generating functions that anti-correlate to the exploration behavior of the algorithm, i.e., larger values of a, b imply a smaller exploration behavior. The circumstance that a, b influence the argument of the sinusoidal functions has a negligible effect on the performance (compare the results for different values of ϕ below). In the case $Q = Q_2$, the parameters a, b also affect the matrix T_d, which is visualized for $a \neq b$ in Figure 4.5 with (a) $a = 2$ and $b = 1$; (b) $a = 1$ and $b = 2$; and (c) $a = 0.5$ and $b = 2$. In case (c) it holds $\mathrm{rk}(T_d) = 2$; otherwise $\mathrm{rk}(T_d) = 4$. All the singular values of W are set to 1. Note that for $a = b$ a behavior similar to that in Figure 4.4 (a) and (b) is attained. For the three different settings, a sequence length of $m = 17$

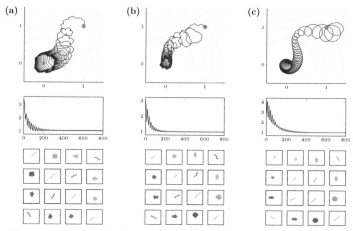

Figure 4.5. An illustration showing how the parameters a, b of the sinusoidal gradient-generating functions affect the McOpt algorithm's behavior. McOpt with parameters as in Algorithm Setup 1 (p. 41) with $Q = Q_2$ and singular values $\sigma = \{1, 1, 1, 1\}$ and $\sigma = \{1, 1\}$ of W is applied to (4.4). The vertical alignment of the figure is as follows: **(Top)** Visualization of $x_k \in \mathbb{R}^2$; **(Center)** Objective value $J(x_k)$ over k; **(Bottom)** Area plot as described in Section 3.3. The parameters a, b are chosen as follows: **Left (a)** $a = 2$, $b = 1$ yields $m = 17$; **Middle (b)** $a = 1$, $b = 2$ yields $m = 17$; **Right (c)** $a = 0.5$, $b = 2$ yields $m = 16$.

for (a) and (b) and $m = 16$ for (c) is obtained. As we can see, the exploration behavior show different characteristics. The weighting of the diagonal blocks of T_d influences the exploration behavior w.r.t. to the distance of x_k to x^* (in case of our test function (4.4)). For example for b), a larger exploration around x_0 than the exploration around x^* can be observed. However, this depends on the choice of f and g and their values at x_0 and x^*. If, for example, $J(x^*) = 0$, this would lead with the given gradient-generating functions in Algorithm Setup 1 (p. 41) to $f(J(x^*)) = 0$ and $g(J(x^*)) = b^{-1/2}$. Thus, choosing a small/large does not influence the behavior in the neighborhood of x^*, whereas choosing b large shrinks the neighborhood of convergence (compare Figure 4.5 (b) and (c) with Figure 4.5 (a)).

Obviously, the parameter ϕ only influences the gradient-generating functions f, g. An illustration of x_k and $J(x_k)$ for (a) $\phi = 0$; (b) $\phi = 3$; and (c) $\phi = 6$, with the default design parameters and $Q = Q_2$ in Algorithm Setup 1 (p. 41) is depicted in Figure 4.6. As we observe, especially in the first steps (marked by red color), ϕ affects the direction of exploration, whereas the convergence behavior is not affected. Thus, in the sequel of this chapter, we always set $\phi = 0$.

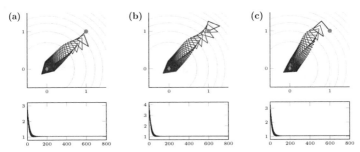

Figure 4.6. An illustration showing how the parameter ϕ of the sinusoidal gradient-generating functions affects the McOpt algorithm's behavior. McOpt with parameters as in Algorithm Setup 1 (p. 41) with $Q = Q_2$ and singular values $\sigma = \{2\sqrt[4]{3}, 2\sqrt[4]{3}\}$ of W is applied to (4.4). The vertical alignment of the figure is as follows: **(Top)** Visualization of $x_k \in \mathbb{R}^2$; **(Bottom)** Objective value $J(x_k)$ over k. The parameter ϕ is chosen as follows: **Left (a)** $\phi = 0$; **Middle (b)** $\phi = 3$; **Right (c)** $\phi = 6$.

Step Size h

On the one hand, the step size accelerates the convergence if it is chosen large, and on the other hand, it shrinks the neighborhood of convergence if it is chosen small. This is depicted for the default parameters as defined in Algorithm Setup 1 (p. 41) with $Q = Q_2$ for (a) $h = 0.25$; (b) $h = 0.1$; and (c) $h = 0.01$ in Figure 4.7. As expected, the largest step size results not only in the fastest convergence speed but also in a large neighborhood of convergence. In contrast, the smallest step size yields a small neighborhood of convergence but a much slower convergence speed. The two advantages can be combined by adaptive step size procedures as discussed in Section 4.4.

To understand the behavior of different gradient-generating functions, we impose the following setup:

Algorithm Setup 2. McOpt with characteristic **[C2]** corresponding to (3.33) with

- Generating functions: $g(z) = -f(z) \int f^{-2}(z) dz$

- $T_d = \begin{bmatrix} 0 & -I \\ I & 0 \end{bmatrix} \in \mathbb{R}^{4 \times 4}$

and singular values $\sigma = \{1, 1, 1, 1\}$ of W. The initial value is set to $x_0 = [1\ 1]^\top$, and the step size is $h = 0.05$.   •

Gradient-generating functions f, g

In the five studies above (Figure 4.3 – Figure 4.7), the gradient-generating functions were based on sinusoidal functions. In the next analysis we consider the following function pairs

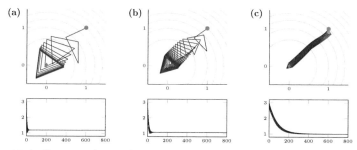

Figure 4.7. An illustration showing how the step size h affects the McOpt algorithm's behavior. McOpt with parameters as in Algorithm Setup 1 (p. 41) with $Q = Q_2$ is applied to (4.4). The vertical alignment of the figure is as follows: **(Top)** Visualization of $x_k \in \mathbb{R}^2$; **(Bottom)** Objective value $J(x_k)$ over k. The step size is chosen as follows: **Left (a)** $h = 0.25$; **Middle (b)** $h = 0.1$; **Right (c)** $h = 0.01$.

(a) $f(z) = 1, g(z) = -z$,

(b) $f(z) = \exp(-z/2), g(z) = \exp(z/2)$,

(c) $f(z) = \sqrt{z}\sin(10\ln(z)), g(z) = \sqrt{z}\cos(10\ln(z))$,

based on Theorem 5. The evolution of x_k and $J(x_k)$ for the different gradient-generating functions specified in (a) – (c) are visualized in Figure 4.8. As we can see in cases (a) and (b), the convergence speed is similar to that in Figure 4.4, where the same singular values are chosen. The difference lies in the exploration behavior, which is more distinct here because of the scaling of the gradient-generating functions by the objective function itself. Case (c) looks very promising; fast practical asymptotic convergence into a small neighborhood of $J(x^*)$. In this view, it might be a bit unfair to compare (c) with the rest of the cases given that in this setting the scaling of the gradient step in the form of the Lie bracket is $[f, g](z) = -10$ (compare with -1 for (a) and (b)); thus, a larger step is performed. However, the gradient scaling depends only on the argument of the sinusoidal function. In a more general form, we have

$$f(z) = \sqrt{z}\sin(\mu\ln(z) + \phi) \quad \text{and} \quad g(z) = \sqrt{z}\cos(\mu\ln(z) + \phi) \tag{4.6}$$

with $\mu \in \mathbb{R}_{>0}$, such that the Lie bracket between f, g results in $[f, g](z) = -\mu$. Another promising and similar class of gradient-generating functions is

$$f(z) = z\sin\left(\frac{\mu}{z} + \phi\right) \quad \text{and} \quad g(z) = -z\cos\left(\frac{\mu}{z} + \phi\right), \tag{4.7}$$

where the Lie bracket between f, g is also $[f, g](z) = -\mu$.

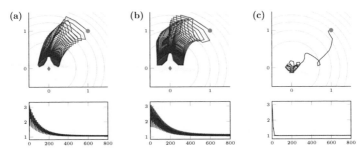

Figure 4.8. An illustration showing how different gradient-generating functions affect the McOpt algorithm's behavior. McOpt with parameters as in Algorithm Setup 2 (p. 46) is applied to (4.4). The vertical alignment of the figure is as follows: **(Top)** Visualization of $x_k \in \mathbb{R}^2$; **(Bottom)** Objective value $J(x_k)$ over k. The gradient-generating functions are chosen as follows: **Left (a)** $f(z) = 1$, $g(z) = -z$; **Middle (b)** $f(z) = \exp(-z/2)$, $g(z) = \exp(z/2)$; **Right (c)** $f(z) = \sqrt{z}\sin(5\ln(z))$, $g(z) = \sqrt{z}\cos(5\ln(z))$.

Characteristic [C1]

The numerical studies above in this section dealt with the algorithm characteristic **[C2]**. As we will observe in the following, the above results also hold for the characteristic **[C1]**. First, we define three setups of our algorithm:

Algorithm Setup 3. McOpt with characteristic **[C1]** corresponding to (3.41) with

- Generating functions: $f(z) = \sin(z)$, $g(z) = \cos(z)$

- $T_d = \begin{bmatrix} -0.5I & -I \\ I & -0.5I \end{bmatrix} \in \mathbb{R}^{4\times4}$

The initial value is set to $x_0 = [1\ 1]^\top$, and the step size is $h = 0.05$.   •

Algorithm Setup 4. McOpt with characteristic **[C1]** corresponding to (3.40) with

- Generating functions: $f(z) = \sqrt{z}\sin((\mu - 1/2)\ln(z))$, $g(z) = \sqrt{z}\cos((\mu - 1/2)\ln(z))$ with $\mu = 20$

- $T_d = \begin{bmatrix} -0.5I & -I \\ I & -0.5I \end{bmatrix} \in \mathbb{R}^{4\times4}$

The initial value is set to $x_0 = [1\ 1]^\top$, and the step size is $h = 0.05$.   •

Algorithm Setup 5. McOpt with characteristic **[C1]** corresponding to (3.40) with

- Generating functions: $f(z) = z\sin\left(-1/2\ln(z^2) + z^{-1}\right)$,
  $g(z) = -z\cos\left(-1/2\ln(z^2) + z^{-1}\right)$

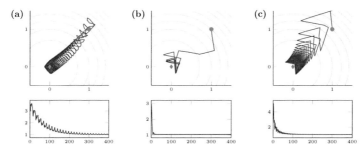

Figure 4.9. An illustration of the characteristic **[C1]** of McOpt applied to (4.4). The vertical alignment of the figure is as follows: **(Top)** Visualization of $x_k \in \mathbb{R}^2$; **(Bottom)** Objective value $J(x_k)$ over k. The design parameters and functions are chosen as follows: **Left (a)** Algorithm Setup 3 (p. 48) yields $m = 14$; **Middle (b)** Algorithm Setup 4 (p. 48) yields $m = 8$; **Right (c)** Algorithm Setup 5 (p. 49) yields $m = 8$.

- $T_d = \begin{bmatrix} Q & -I \\ I & Q \end{bmatrix} \in \mathbb{R}^{4 \times 4}$ with $Q = \begin{bmatrix} -1.1 & 0 \\ 0 & -0.9 \end{bmatrix}$

The initial value is set to $x_0 = \begin{bmatrix} 1 & 1 \end{bmatrix}^\top$, and the step size is $h = 0.05$.

The numerical results of Algorithm Setup 3 – Algorithm Setup 5 (p. 48) applied to (4.4) are visualized in Figure 4.9. As we observe, Figure 4.9 (a) and Figure 4.4 (a) reveal a similar behavior, although for the current setting only one function evaluation per iteration is necessary. The gradient-generating function in Algorithm Setup 4 (p. 48) and Algorithm Setup 5 (p. 49) are the counterparts to (4.6) and (4.7), respectively. In both settings, scaled negative gradients of the objective J are approximated; thus, a faster convergence can be observed.

Compared with the characteristic **[C2]**, the choices of the design parameters and functions are a bit more limited, but for similar settings the same behavior can be expected, although m could be larger for **[C1]** because of limitations in the choices of T_d—in exchange, only one function evaluation per iteration is required. Moreover, the degree of freedom in choosing the singular values of W is restricted to pair-wise identical singular values and one specific case of T_d (cf. Corollary 2).

Conclusion and Discussion

We presented a qualitative study regarding each design parameter and function of McOpt. In essence, through these investigations we gained a qualitative numerical understanding in terms of the convergence and exploration behavior. Consequently, we can conclude that a good starting point in choosing the algorithm's design parameters and functions is the basic setting as presented in (3.34) with $a = b = 1$ and a small step size, e.g. $h = 0.01$. The choice of gradient-generating functions is the next step in tuning the algorithm's parameters, e.g. choosing (4.6) or (4.7) (note that T_d must be adapted). Following on that, we choose the singular values in the same scale, but a bit smaller than the ones which reveal a minimal m w.r.t. T_d. Eventually, T_d in combination with the singular values can be utilized to scale each sequence $e_i^\top \{w_k\}_{k=0}^{m-1}$ for $i = 1, \dots, 2n$. This can be helpful if some knowledge of the objective is present and a larger exploration in some dimension is intended.

4.3 Design Parameters and Functions: Quantitative Study

The objective of this section is a quantitative study of the algorithm's design parameters and functions, specifically gradient-generating functions f, g, matrix T_d, and singular values $\{\sigma_i\}_{i=0}^{\mathrm{rk}(T_d)}$ of W. We study permutations of various parameter settings and evaluate the algorithm's numerical results on a quantitative level, i.e., evaluating convergence speed and the average gradient approximation error as performance measurements. At first, we define the performance measurement metrics, followed by the description of the parameter study, including algorithm setups, involved parameters, and analyzed number of data points. Eventually we discuss the results.

Test Function and Performance Evaluation

The quantitative design parameter and function study is also carried out on the two-dimensional spherical test function (4.4) with global minima $x^* = 0 \in \mathbb{R}^2$ and $J(x^*) = 1$. As in the previous qualitative study, we limit ourselves on this rather simple test function for the sake of comparability and provide a benchmarking study of more challenging objective functions in Section 4.5. For the quantitative evaluation of the algorithm's numerical results, we introduce the following setting and performance measurements:

- McOpt is applied to (4.4) with each parameter setting (defined in next paragraph) and the same set of 10 initial values x_0, where $x_0 \in \{x \in \mathbb{R}^2 : 0.5 \leq \|x - x^*\|_2 \leq 2\}$.

- Convergence at step k^* into a neighborhood of x^*, i.e., $x_k \in \mathcal{U}_{\sqrt{h}}(x^*)$, where $k^* = \arg\min_{k \in \mathbb{N}_0} \{\|x_k - x^*\|_2 \leq \sqrt{h}\}$.

- Averaged gradient approximation error $\rho = {}^1/10 \sum_{k=0}^{9} \|x_{m+k} - (x_k - h\nabla J(x_k))\|_2$, i.e., for (4.4) it holds that $\rho = {}^1/10 \sum_{k=0}^{9} \|x_{m+k} - x_k + 2hx_k\|_2$.

Note that we apply a convergence test based on a fixed area of convergence in contrast to the method as stated in equation (2.2) in Moré and Wild (2009). Therein, k^* is defined by $k^* = \arg\min_{k \in \mathbb{N}_0} \{J(x_k) \leq J(x^*) + \tau(J(x_0) - J(x^*))\}$ for $\tau \in \mathbb{R}_{>0}$. Because of the fixed step size, the neighborhood of convergence depends on h; therefore, we limit our convergence test in this section to the proposed analysis. However, in Section 4.5 we apply performance measurements as in Moré and Wild (2009), because the algorithm class McOpt is equipped with tuning rules, including decreasing step sizes as presented in Section 4.4.

The averaged gradient approximation error ρ is the average of the first ten second-order (Lagrange) remainder terms gained by applying *Taylor's Theorem* on m evolutionary steps of the right hand side of (4.1) or (4.2) as executed in Lemma 6 and applied in Theorem 1. The gradient term $h\nabla J(x_k)$ must dominate this remainder term to ensure a decreasing step (cf. Theorem 2). However, low values of ρ are not necessary or may not even lead to low values of k^*, as we will observe.

Parameter Study

In the view of Theorem 5 and Theorem 6, various parameter settings are available. Conceptually, in the case of **[C2]** we have seven main classes of matrices T_d (see (3.33) – (3.39)) including numerous gradient-generating functions f, g with multiple parameter options, such as constants $a, b \in \mathbb{R}_{>0}$, $c, \phi \in \mathbb{R}$, and the skew-symmetric matrix $Q \in \mathbb{R}^{n \times n}$ (cf. Theorem 5). In the case of **[C1]** we have two main classes of matrices T_d (see (3.40) – (3.41)) including different gradient-generating functions with multiple parameter options, such as function $r : \mathbb{R} \to \mathbb{R}$, normal matrix $Q \in \mathbb{R}^{n \times n}$ (with $Q + Q^\top$ negative definite), and constants $a \in \mathbb{R} \setminus \{0\}$, $b \in \mathbb{R}_{>0}$, and $\phi \in \mathbb{R}$ (cf. Theorem 6). As concluded in Section 4.2, the parameter ϕ has no effect on the convergence behavior; thus, we neglect it in the sequel of this study and set $\phi = 0$. All considered algorithm setups, including choices of parameters, are listed in Appendix C.1.1 as Algorithm Setup 10 – Algorithm Setup 19. Every setup is tested on the number of convergence steps k^* and the averaged gradient approximation error ρ. The generated simulation data is quantitatively analyzed as presented exemplarily in Figure 4.10 and is discussed in the next paragraph. All additional illustrations of the data (in total 4255 data points) can be found in Appendix C.1.2. To discuss the results, we introduce a color code for the discrete heat maps as illustrated in Figure 4.10 and provided in the corresponding figures in Appendix C.1.2. In particular, a colorbar with ten colors (green to black) is used to identify promising parameter setting. In terms of k^* we set a parameter setting to green (■) if the algorithm with this setting for all initial points x_0 converges in average to $\mathcal{U}_{\sqrt{h}}(x^*)$ in 0 to 100 steps. We set it to neon green (□) if convergence into $\mathcal{U}_{\sqrt{h}}(x^*)$ in 101 to 200 steps is evaluated, and so forth, as depicted in the legend of Figure 4.10. A similar color coding is applied to the averaged gradient error ρ. Note that the size of ρ is not relevant because it depends on J. Conceptually, every permutation of design parameters and functions of the presented McOpt algorithm class given by Algorithm Setup 10 – Algorithm Setup 19 in Appendix C.1.1 is applied to the optimization objective (4.4) for a maximum of 1000 iterations and ten initial points x_0. Then the mean value of the performance measurements k^* and ρ is taken for each parameter setting over the set of initial points. If the convergence criteria for $k \leq 1000$ steps of one initial point x_0 is not fulfilled, we set $k^* = \infty$ (■) for the tested parameter setting. We want to emphasize that the focus of this study lies on the best parameter setting with a fixed step size $h = 0.05$, whereas, in the sequel, our target is to improve these settings by step size and exploration sequence tuning (cf. Section 4.4) to improve k^* and ρ.

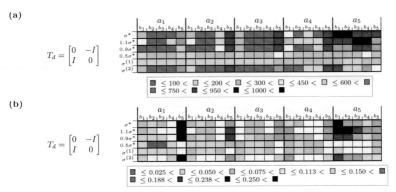

Figure 4.10. Visualization of the data points for Algorithm Setup 10 (see Appendix C.1.1), which corresponds to the triples (T_d, f, g) in (3.34). For the sake of completeness (explained in detail in Appendix C.1.1), $a, b \in \{0.1, 0.5, 1, 2, 10\}$, σ^* are the singular values of W such that the minimal number of steps m w.r.t. T_d are revealed, $\sigma^{(1)} = \{1, \ldots, 1\}$, and $\sigma^{(2)} = \{2, \ldots, 2\}$. **Top (a)** Steps k^* to converge into $\mathcal{U}_{\sqrt{h}}(x^*)$; **Bottom (b)** Averaged gradient approximation error ρ.

Discussion

All observations in this paragraph are based on the data point visualizations Figure C.1 – Figure C.18 in Appendix C.1.2, similar to Figure 4.10. The conclusions of this parameter study are related to the triples (T_d, f, g) in Theorem 5 and Theorem 6.

Obviously, if the singular values of W are chosen such that the minimal sequence length m w.r.t. T_d is revealed, given by σ^* (see Appendix D for the construction of W), this leads to a smaller k^* in general. However, choosing the singular values of W a bit smaller yields a slightly better behavior in terms of convergence speed. Moreover, the averaged gradient approximation error is smaller; ρ correlates to the singular values of W, i.e. small singular values yield a smaller averaged gradient approximation error but a larger sequence length m. Thus, a faster convergence and hereby less function evaluations are gained by a larger uncertainty in the gradient approximations.

In terms of the scaling parameters a, b, values between 0.5 and 2 reveal a good performance for sinusoidal generating functions. In the cases of (3.35) – (3.37), these parameters are more sensitive and must be chosen a magnitude smaller. Moreover, and because of a symmetric test function, $a = b$ should be the first choice. These parameters also influence the block-diagonal elements of T_d for some triples (T_d, f, g), where we observed the algorithm's sensitive and poor behavior. Thus, it should be ensured that the entries of T_d are close to one.

The "best" behavior in terms of convergence speed in this study corresponds to (3.33) and (3.40) with gradient-generating functions

$$f(z) = a^{-1/2}\sqrt{z}\sin(\mu\sqrt{ab}\ln(z)) \quad \text{and}$$
$$g(z) = b^{-1/2}\sqrt{z}\cos(\mu\sqrt{ab}\ln(z)), \tag{4.8a}$$

$$f(z) = a^{-1/2}z\sin(\mu\sqrt{ab}z^{-1}) \quad \text{and}$$
$$g(z) = b^{-1/2}z\cos(\mu\sqrt{ab}z^{-1}), \tag{4.8b}$$

similar to (4.6) and (4.7) with $a, b, \mu \in \mathbb{R}_{>0}$ for characteristic [C2], and

$$f(z) = b^{-1/2}\sqrt{z}\sin(a/2\ln(b^{-1}z) + b\mu\ln(z)) \quad \text{and}$$
$$g(z) = b^{-1/2}\sqrt{z}\cos(a/2\ln(b^{-1}z) + b\mu\ln(z)), \tag{4.9a}$$

$$f(z) = b^{-1/2}z\sin(-a/2\ln(b^{-1}z) + b\mu z^{-1}) \quad \text{and}$$
$$g(z) = -b^{-1/2}z\cos(-a/2\ln(b^{-1}z) + b\mu z^{-1}), \tag{4.9b}$$

similar to Algorithm Setup 4 (p. 48) and Algorithm Setup 5 (p. 49) with $-a, b, \mu \in \mathbb{R}_{>0}$ for characteristic [C1]. As mentioned in the prior section, (4.8) and (4.9) reveal a scaled gradient step by μ. Clearly, this accelerates the convergence speed, although the gradient error is larger (note that it is compared according to a scaled gradient step; cf. Figure C.9 – Figure C.12 and Figure C.15 – Figure C.18 in Appendix C.1.2).

Moreover, (3.34) and (3.39) show very promising results (cf. Figure C.1 and Figure C.7 in Appendix C.1.2). Especially if the objective is very steep, these generating functions are a good starting point because of their bounded (sinusoidal) appearance.

In summary, in this study we presented the sensitivity and choice of the design parameters and functions and suggested promising generating functions for the McOpt algorithm class. Nevertheless, this is based on the two-dimensional spherical test function and must be seen as a starting point for choosing the design parameters and functions. Therefore, in Section 4.5 we run a selection of algorithm settings based on our conclusions of this section on various challenging benchmarking test functions and compare it with other derivative-free optimization algorithms of the same class, i.e., with gradient approximating procedures.

4.4 Performance Tuning

In this section the presented McOpt algorithm class is extended by a variable and adaptive step size and a so-called restarting of the exploration sequence to gain performance improvements. First, several results are presented and then evaluated on the two dimensional spherical test function (4.4). We consider the following three McOpt setups:

Algorithm Setup 6. McOpt with characteristic [C2] corresponding to (3.34) with

- Generating functions: $f(z) = \sin(z)$, $g(z) = \cos(z)$

- $T_d = \begin{bmatrix} Q & -I \\ I & Q \end{bmatrix} \in \mathbb{R}^{4\times4}$ with $Q = \begin{bmatrix} 0 & -1 \\ 1 & 0 \end{bmatrix}$

The singular values of W are set to $\sigma = \{1, 1\}$, and the initial value to $x_0 = [1\ 1]^\top$. •

Algorithm Setup 7. McOpt with characteristic **[C1]** corresponding to (3.41) with

- Generating functions: $f(z) = \sin(z)$, $g(z) = \cos(z)$

- $T_d = \begin{bmatrix} -0.5I & -I \\ I & -0.5I \end{bmatrix} \in \mathbb{R}^{4\times4}$

The singular values of W are $\sigma = \{1, 1, 1, 1\}$, and the initial value is set to $x_0 = [1\ 1]^\top$. •

Algorithm Setup 8. McOpt with characteristic **[C2]** corresponding to (3.33) with

- Generating functions: $f(z) = \sqrt{z}\sin(5\ln(z))$, $g(z) = \sqrt{z}\cos(5\ln(z))$

- $T_d = \begin{bmatrix} 0 & -I \\ I & 0 \end{bmatrix} \in \mathbb{R}^{4\times4}$

The singular values of W are set to $\sigma = \{0.6, 0.6, 0.6, 0.6\}$, and the initial value to $x_0 = [1\ 1]^\top$.

•

For a fixed step size $h = 0.05$, the evolution of x_k and $J(x_k)$ for Algorithm Setup 6 – Algorithm Setup 8 with $m = 13$, $m = 14$, and $m = 36$, respectively, are depicted in Figure 4.11. The three setups are chosen based on their good performance and results of the qualitative and quantitaive study (cf. Section 4.2 and Section 4.3). Moreover, both characteristics **[C1]** and **[C2]** are present, as well as a scaled gradient version.

4.4.1 Step Size

The algorithms (4.1) and (4.2) suffer in convergence speed for small step sizes and show large oscillations in a neighborhood of convergence for large step sizes (see Figure 4.7). From a practical and implementation perspective it is straightforward to investigate decreasing and adaptive step sizes to achieve faster convergence, as well as a good steady state behavior, i.e., a small neighborhood of convergence around a local minima x^*.

Decreasing Step Size

As indicated in Section 3.2 and Theorem 3, including its proof in Appendix B.3, the neighborhood of convergence scales with $h^{3/2}$. In this view, a small h or a variable step size with $h_k \to 0$ for $k \to \infty$ yields a smaller or vanishing neighborhood of convergence, respectively. Therefore, asymptotic convergence instead of practical asymptotic convergence can be achieved (cf. Theorem 2 and Theorem 3). Additionally, a large step size for the first

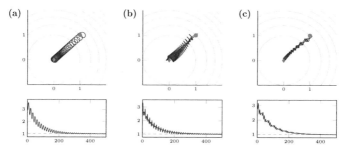

Figure 4.11. Three different settings are utilized for comparison of various performance tuning rules for the McOpt algorithm class with $h = 0.05$ applied to (4.4). The vertical alignment of the figure is as follows: **(Top)** Visualization of $x_k \in \mathbb{R}^2$; **(Bottom)** Objective value $J(x_k)$ over k. The design parameters and functions are chosen as follows: **Left (a)** Algorithm Setup 6 (p. 54); **Middle (b)** Algorithm Setup 7 (p. 54); **Right (c)** Algorithm Setup 8 (p. 54).

steps allow a larger exploration of the objective and a larger gradient step; thus, faster convergence towards a local minima. Conceptually, and as introduced in Theorem 3, a variable decreasing step size h_k, but constant over a period of length m as stated in (3.18), e.g. $h_k = h_0(\lfloor k/m \rfloor + 1)^{-1}$ leads to a semi-global asymptotic convergence result. Thus, both advantages—fast convergence and a small neighborhood of convergence—are combined by this step size rule. The performance improvement is visualized in Figure 4.12 with $h_0 = 1$. As is apparent, the convergence speed is accelerated by a factor larger than ten in this simulation example compared with Figure 4.11. Moreover, for k large it holds that $x_k \to x^*$; thus, asymptotic convergence to the minimum x^*. In essence, smaller singular values can be chosen to achieve a more accurate gradient approximation (cf. Section 4.3); this leads to a larger sequence length m that is balanced by larger steps because of larger step sizes h, while the gradient approximation error is reduced.

Adaptive Step Size

The decreasing step size procedure in the paragraph above ensures a vanishing neighborhood of convergence. However, initial points x_0 could already be sensitive with respect to large step sizes h_k (e.g. steep functions), but small initial step sizes lead to a slow convergence speed. In this view, a similar approach as the well-known adaptive Runge-Kutta procedure (cf. Cheney and Kincaid (2012), Chapter 7.3, p. 320 – 328) is developed for McOpt in the sequel. For (4.2), i.e, characteristic **[C2]**, the following step size rule for every $k \in \mathbb{N}$

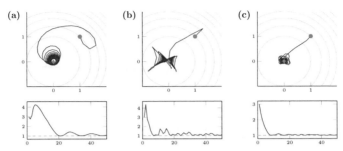

Figure 4.12. An illustration of the decreasing step size rule (3.18) for McOpt with initial step size $h_0 = 1$ and $h_k = h_0(\lfloor k/m \rfloor + 1)^{-1}$ applied to (4.4). The vertical alignment of the figure is as follows: **(Top)** Visualization of $x_k \in \mathbb{R}^2$; **(Bottom)** Objective value $J(x_k)$ over k. The design parameters and functions are chosen as follows: **Left (a)** Algorithm Setup 6 (p. 54); **Middle (b)** Algorithm Setup 7 (p. 54); **Right (c)** Algorithm Setup 8 (p. 54).

with initial step size $h_0 \in \mathbb{R}_{>0}$ such that

$$e_k = \|\hat{x}_k - x_k\|_2$$

$$= \frac{1}{2} \left\| \Big(Y\big(f(J(\hat{x}_k)), g(J(\hat{x}_k))\big) - Y\big(f(J(x_k)), g(J(x_k))\big) \Big) W^* e_{k \bmod(m)+1} \right\|_2 \qquad (4.10a)$$

$$h_{k+1} = \begin{cases} v_1 h_k & e_k < \epsilon_1 \\ h_k & \epsilon_1 < e_k < \epsilon_2 \\ v_2 h_k & \epsilon_2 < e_k \end{cases} \qquad (4.10b)$$

where $\epsilon_1, \epsilon_2 \in \mathbb{R}_{>0}$, $v_1 \in \mathbb{R}_{>1}$, and $0 < v_2 < 1$ is applied. Although, the step size h_k is *not* constant over a period of length m, compared with the decreasing step size rule (3.18), this tuning heuristic works very well in practice. Note that if v_1, v_2 are chosen close enough to one, Lemma 3 still holds. From a conceptual perspective, (4.10) calculates the distances between $\hat{x}_k$ and x_{k+1} in (4.2) and reduces or increases the step size if this distance is large or small, respectively. The distance depends, for example, on the steepness, condition number, or smoothness of the objective.

Nevertheless, the steady state amplitude depends on ϵ_1, ϵ_2. Thus, if $\|x_k - x_{k-m}\|_2 < \epsilon_3 \sqrt{h_k}$ for every $k \in \mathbb{N}$ for some $\epsilon_3 \in \mathbb{R}_{>0}$, we apply

$$\epsilon_1 = v_3 \epsilon_1, \qquad \epsilon_2 = v_3 \epsilon_2 \qquad (4.11)$$

with $0 < v_3 < 1$. Therefore, if the difference of x_k over one period length m is small, ϵ_1, ϵ_2 are reduced such that the interval for decreasing the step size, i.e., $e_k > \epsilon_2$ in (4.10) is diminished and h_k is becoming more sensitive w.r.t. e_k. The performance improvement by (4.10) with (4.11) is visualized in Figure 4.13 for $h_0 = 0.001$, $\epsilon_1 = \epsilon_3 = 0.1$, $\epsilon_2 = 0.2$, $v_1 = 1.1$, $v_2 = 0.9$, and $v_3 = 0.8$. Note that (4.10) can not be applied to Algorithm Setup 7 because it is designed

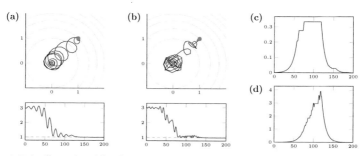

Figure 4.13. An illustration of the adaptive step size rule (4.10) with (4.11) for McOpt with $h_0 = 0.001$, $\epsilon_1 = \epsilon_3 = 0.1$, $\epsilon_2 = 0.2$, $v_1 = 1.1$, $v_2 = 0.9$, and $v_3 = 0.8$ applied to (4.4). The vertical alignment of the figure is as follows: **Top (a) and (b)** Visualization of $x_k \in \mathbb{R}^2$; **Bottom (a) and (b)** Objective value $J(x_k)$ over k. The design parameters and functions are chosen as follows: **Left (a)** Algorithm Setup 6 (p. 54); **Middle (b)** Algorithm Setup 8 (p. 54); adaptive step size h_k for **Right top (c)** Algorithm Setup 6 (p. 54) and **Right bottom (d)** Algorithm Setup 8 (p. 54).

solely for algorithm characteristic **[C2]**. As we can see in Figure 4.13, the evolution of x_k begins with a small exploration because of $h_0 = 0.001$. This is enlarged by (4.10) such that the convergence is accelerated; while approaching x^*, the step size is reduced again because of (4.11). The main advantage of this tuning heuristic is a small initial step size—such that convergence is ensured—which is adapted based on the objective function.

4.4.2 Exploration Sequence

Based on the periodic and explorative nature of the presented McOpt algorithm class, $J(x_{k+1}) > J(x_k)$ for $k \geq 0$ is not excluded in the algorithm's behavior. Because m steps are utilized to approximate first-order information of the objective, larger values can be visited to receive this information. However, this explorative behavior can be tackled more efficiently for performance tuning, as presented in the next two paragraphs.

Restarting of Exploration Sequence

As illustrated in Figure 4.14 (a), an in average decreasing objective value $J(x_k)$ over iterations k is a common behavior for the standard versions of the McOpt algorithm class. The main idea is that for every $k \in \mathbb{N}$ with $k \bmod (m) \equiv 0$ a new initial point x_{k-1}^0 such that

$$x_{k-1}^0 = \min\{x_\ell\}_{\ell=k-m}^{k-1} \tag{4.12}$$

is set, i.e., every m steps we restart at the best visited point over one period as depicted in Figure 4.14 (a). Conceptually, the algorithm is applied for m steps over and over again. Thus, the convergence result of Theorem 2 is still valid because an approximated gradient step or a step which leads to a smaller function value is performed, i.e., $J(x_{k+m}) - J(x_k) \leq 0$.

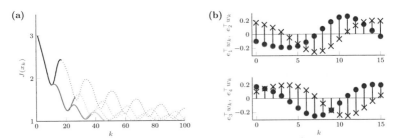

Figure 4.14. Illustrations of the main idea to utilize the exploratory behavior of the McOpt algorithm class to accelerate convergence. **Left (a)** Restarting at the best visited point w.r.t. the lowest function value. Bold lines represent the first $m = 16$ steps of each restarting to approximate the gradient of J. **Right (b)** Periodic exploration sequence $\{w_k\}_{k=0}^{15}$, where $e_1^\top w_k$ and $e_3^\top w_k$ (•), and $e_2^\top w_k$ and $e_4^\top w_k$ (×), respectively. The idea is to restart with some random index $p \in \{0, \ldots, m-1\}$ such that $We_{(p+k)\,\mathrm{mod}(m)+1}$ is applied to (4.1) or (4.2) for $k \in \mathbb{N}_{\geq 0}$.

Randomized Restart of Exploration Sequence

For objectives such as the spherical test function (4.4), which are symmetric and smooth, restarting reveals only a minor improvement in convergence speed, given that the trajectory of each period is almost identical in its exploration shape. Therefore, a randomized restarting of the exploration sequence is introduced. Because of the periodicity of the exploration sequence, as depicted in Figure 4.14 (b), one can apply the following: after each gradient approximation cycle, i.e., every m steps, we re-index the exploration sequence matrix, i.e., choose randomly an index $p \in \{0, \ldots, m-1\}$ and re-structure the exploration sequence matrix such that $W = [w_p\; w_{p+1}\; \cdots w_{m-1}\; w_0\; \cdots w_{p-1}]$ holds. In other words, $We_{(p+k)\,\mathrm{mod}(m)+1}$ is applied to (4.1) or (4.2) for $k \in \mathbb{N}_0$. This rather simple tuning rule can have a beneficial impact on the convergence speed as visualized for Algorithm Setup 6 – Algorithm Setup 8 (p. 54) in Figure 4.15. Compared with Figure 4.11, this is a factor increase of two in terms of convergence speed.

4.4.3 Discussion

We presented two methods to accelerate convergence speed and shrink the neighborhood of convergence. First, a variable and adaptive step size h_k. Second, we applied the exploration sequence in a random fashion while restarting the algorithm every m steps. In both cases a gain in convergence speed was achieved, as we can observe by comparing Figure 4.11 with Figure 4.12, Figure 4.13, and Figure 4.15.

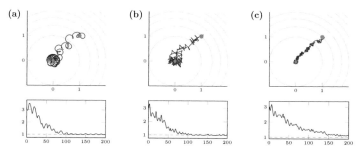

Figure 4.15. Restarting the algorithm every m steps with (4.12) for McOpt with $h = 0.05$ applied to (4.4). The vertical alignment of the figure is as follows: **(Top)** Visualization of $x_k \in \mathbb{R}^2$; **(Bottom)** Objective value $J(x_k)$ over k. The design parameters and functions are chosen as follows: **Left (a)** Algorithm Setup 6; **Middle (b)** Algorithm Setup 7; **Right (c)** Algorithm Setup 8.

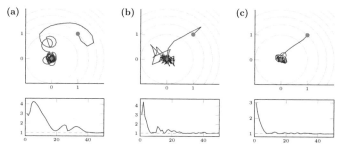

Figure 4.16. Decreasing step size rule (3.18) combined with a random restarting policy of the exploration sequence W for McOpt with initial step size $h_0 = 1$ and $h_k = h_0(3\lfloor k/m \rfloor + 1)^{-1}$ applied to (4.4). The vertical alignment of the figure is as follows: **(Top)** Visualization of $x_k \in \mathbb{R}^2$; **(Bottom)** Objective value $J(x_k)$ over k. The design parameters and functions are chosen as follows: **Left (a)** Algorithm Setup 6; **Middle (b)** Algorithm Setup 7; **Right (c)** Algorithm Setup 8.

Eventually, combining a decreasing step size as in (3.18) and a randomized restarting policy of the exploration sequence with new initial points (4.12) seems very promising. This is visualized in Figure 4.16. Note that only a minor improvement over Figure 4.12 can be observed. However, we apply this combination of tuning rules in the next section, where we run several benchmarking tests on challenging multi-variable, ill-conditioned, non-convex, and discontinuous test functions.

4.5 Benchmarking

To evaluate the performance and show the practicality of the presented McOpt algorithm class, we test four different settings of McOpt against two well known gradient approximation schemes—simultaneous perturbation stochastic approximation (SPSA; see Spall (1992)) and finite difference stochastic approximation (FDSA; see Blum (1954))—as already introduced in Section 2.1. Additionally, the *Gradientless Descent* (GLD) algorithm by Golovin et al. (2019) is used for comparison. It also belongs to the class of "local explorative" optimization algorithms, although no gradient is approximated, i.e, a (local) direct search procedure is deployed. Moreover, as a benchmark we apply the most simplest zeroth-order optimization scheme, random search. Note that those comparison algorithms are implemented to solve various optimization applications; see, for example, J. Bergstra and Bengio (2012); Kushner and Clark (2012); Golovin et al. (2019); Dong and Chen (2012a); Dong and Chen (2012b); Ko, Lee, and Kim (2008); and Burnett (2003); to name only a few. Moreover, there exist several extensions of those algorithms which are adopted and tailored to specific problems. For comparison, we implement the standard versions of each algorithm, given that the considered optimization problems in this benchmarking study belong to different problem (function) classes.

The aforementioned algorithms and their parameter choices are described in detail by Algorithm Setup 20 – Algorithm Setup 27 in Appendix C.2. The four McOpt algorithm setups—two of each characteristic **[C1]** and **[C2]**—correspond to (3.33), (3.34), (3.40), and (3.41). Our choice of those setups is supported by the discussion and the promising performance achievements in the sections above (Section 4.2 – Section 4.4). Moreover, the McOpt algorithms are equipped with a decreasing step size and randomized restarting of the exploration sequence, as introduced in Section 4.4. We want to underline that we choose the parameters of the algorithms to basic values, e.g. all singular values of W are set to one or such that the minimal sequence length m w.r.t. the given setting is revealed. Hence, we do not perform a tuning of the design parameters. This highlights that the presented algorithm class is not sensitive with respect to the parameter choices, as we will see in the sequel. In contrast, this is a drawback of the compared algorithms SPSA (cf. page 165 – 166 in Spall (2005)) and FDSA. Therefore, we tuned their parameters in advance but set the same parameters for all considered optimization problems such that we have a common ground of comparison. Note that the initial step size of SPSA and FDSA is a magnitude smaller than McOpt, because of observed divergent behaviors for large step sizes, but the slope of the decreasing function of the step size is lower. Hence, the sum over the step size sequence is comparable.

The algorithms in Algorithm Setup 20 – Algorithm Setup 27 (see Appendix C.2.1) are applied to eight n-dimensional objective functions J for $n = 2, 4, 10$, which are proven to be challenging for derivative-free optimization algorithms (see, e.g., Moré and Wild (2009); Hansen et al. (2009); and Golovin et al. (2017)), namely the *Sphere, Branin, Ellipsoid, Styblinski, Rosenbrock, Six Hump Camel, Schwefel*, and *Manevich* functions as depicted in Figure 4.17 for $n = 2$ and stated in Table C.1 (see Appendix C.2). They are far from convex and smooth, many exhibiting high conditioning, multi-modal valleys, and weak global

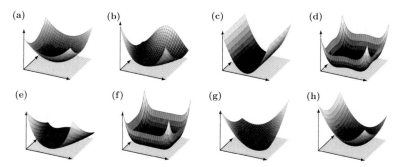

Figure 4.17. Visualization of the test functions for $n = 2$, utilized to benchmark the algorithm class McOpt against simultaneous perturbation stochastic approximation (SPSA), finite difference stochastic approximation (FDSA), gradientless descent (GLD), and random search (RS) (see Appendix C.2.1). The functions are presented in Table C.1 in Appendix C.2: **(a)** Sphere; **(b)** Branin; **(c)** Ellipsoid; **(d)** Styblinski; **(e)** Rosenbrock; **(f)** Six Hump Camel; **(g)** Schwefel; **(h)** Manevich.

structure. Additionally, we introduce three variations of the objective functions. First is *multiplicative noise* (see equation (4.2) – (4.4) in Moré and Wild (2009)), such that

$$J_{\text{mult}}(x) = (1 + 0.5\phi(x))J(x) \quad \text{with} \tag{4.13a}$$

$$\phi(x) = \phi_0(x)\big(4\phi_0(x)^2 - 3\big) \quad \text{where} \tag{4.13b}$$

$$\phi_0(x) = 0.9\sin(100\|x\|_1)\cos(100\|x\|_\infty) + 0.1\cos(\|x\|_2), \tag{4.13c}$$

where $\phi(\cdot)$ is the cubic Chebyshev polynomial. Second is *additive Gaussian noise*, such that

$$J_{\text{add}}(x) = J(x) + \exp(\mathcal{N}(0,1)), \tag{4.14}$$

where $\mathcal{N}(0,1)$ denotes a real-valued random variable drawn from the Gaussian (normal) distribution with mean value zero and standard deviation one. Last is *discontinuous*, such that

$$J_{dc}(x) = J(0.1\,\text{round}(10x)) \tag{4.15}$$

where the objective function consists of piecewise constant polytopes with edge length 0.1. Note that the operator round($\cdot$) denotes the rounding of a real number to its closest integer; element-wise executed on a vector. The undisturbed (non-noisy and continuous) setting is named *normal* and denoted by J. The considered variations J, J_{mult}, J_{add}, and J_{dc} are depicted for the sphere test function (Figure 4.17 (a)) and $n = 2$ in Figure 4.18.

The considered test functions are simple to calculate, and the optimal points $x^* \in \mathbb{R}^n$ are known. Thus, the performance evaluation, in the form of the *normalized logarithmic optimality*

(a)   (b)   (c)   (d)

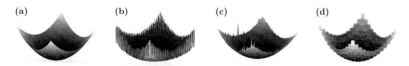

Figure 4.18. Visualization of the test function variations applied to the functions presented in Table C.1 in Appendix C.2 to benchmark the algorithm class McOpt. This is illustrated by the sphere function and its multiplicative noise, additive Gaussian noise, and discontinuous variations for $n = 2$: **(a)** Normal (J); **(b)** Multiplicative noise (J_{mult}); **(c)** Additive Gaussian noise (J_{add}); **(d)** Discontinuous (J_{dc}).

gap $\tau : \mathbb{R} \to \mathbb{R}$, defined by

$$\tau(J(x_k)) := \log\left(\frac{J(x_k) - J(x^*)}{J(x_0) - J(x^*)}\right), \qquad k \geq 0 \tag{4.16}$$

can be easily calculated. Conceptually, (4.16) gives the success rate and competence of the algorithms behavior with respect to the initial values (cf. Moré and Wild (2009)). Specifically, τ represents the relative decrease from the initial point $J(x_0)$. Moreover, τ is invariant w.r.t. the affine transformation $J \mapsto \alpha J + \beta$ with $\alpha \in \mathbb{R}_{>0}$ and $\beta \in \mathbb{R}$. For comparison and performance evaluation, the optimality gap $\tau(J(x_t))$ against the number of function evaluations, where x_t is the best point observed so far after t function evaluations, is considered in the sequel. Thus,

$$\tau(J(x_t)) = \min_{k \in [0, \lfloor t/T \rfloor]} \{\tau(J(x_k))\}, \tag{4.17}$$

with $T \in \mathbb{N}$ function evaluation per iteration. Note that $T = 1$ and $T = 2$ for the characteristics **[C1]** and **[C2]**, respectively, whereas for SPSA, FDSA, GLD, and RS it holds that $T = 2$, $T = 2n$, $T = \log(n^{3/2}) + 10$, and $T = 1$. The benchmarking study is conducted such that every algorithm runs with a computational budget of 5000 function evaluations. For each dimension n, we ran each optimization algorithm on each benchmark test function 20 times, i.e., we initialize each run with 20 different initial values with $\|x_0\|_2 \leq n$ (initial value set is the same for all algorithms and function; only varies with n) and average the intermediate results over the multiple runs.

As examples, the normalized logarithmic optimality gaps for the *Branin* test function for $n = 4$, including its noisy and discontinuous variations, are depicted in Figure 4.19. Note that the filled markers belong to the McOpt algorithm class and the non-filled markers to the comparison algorithms. In this case, we can observe that the McOpt algorithms reveal a competitive and strong performance. An optimality gap of $\tau^* = 10^{-4}$ is achieved in every setting in less or a similar amount of function evaluations in comparison with the benchmarking algorithms SPSA, FDSA, GLD, and RS.

To avoid a case-by-case analysis for every test function and its variations (in total 96 illustrations) we introduce the so-called *data profiles* (cf. Moré and Wild (2009)); nevertheless,

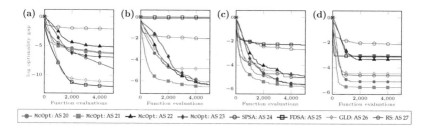

Figure 4.19. Logarithmic optimality gap τ (see (4.17)) for Algorithm Setup 20 – Algorithm Setup 27 (AS 20 – AS 27) as described in Appendix C.2 for the *Branin* test function for $n = 4$ with **Left (a)** Normal (J); **Middle left (b)** Multiplicative noise (J_{mult}); **Middle right (c)** Additive Gaussian noise (J_{add}); **Right (d)** Discontinuous (J_{dc}).

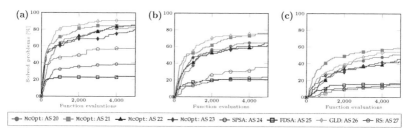

Figure 4.20. Data profiles for Algorithm Setup 20 – Algorithm Setup 27 (AS 20 – AS 27) as described in Appendix C.2 for the test functions and its variations as described in Table C.1. **Left (a)** $\tau^* = 10^{-2}$; **Middle (b)** $\tau^* = 10^{-3}$; **Right (c)** $\tau^* = 10^{-4}$.

the logarithmic optimality gap figures—similar to Figure 4.19—are presented for every test function and its variations for $n = 2, 4, 10$ by Figure C.19 – Figure C.26 in Appendix C.2. The data profiles present the percentage of considered problems—eight objectives in four variations for three dimensions (96 problems)—which achieve an optimality gap τ^* by the given algorithms in t function evaluations. Therefore, it can be seen as a condensed visualization of the optimality gap illustrations in Figure 4.19 and Figure C.19 – Figure C.26 in Appendix C.2. Three data profiles for $\tau^* = 10^{-2}$, $\tau^* = 10^{-3}$, and $\tau^* = 10^{-4}$ are depicted in Figure 4.20 for all 96 considered problems. Additionally, in Figure 4.21, an illustration of the separated objectives J, J_{mult}, J_{add}, and J_{dc} for $\tau^* = 10^{-3}$ is presented, i.e., 24 problems per data profile.

The data profiles in Figure 4.20 and Figure 4.21 and the related logarithmic optimality gap illustrations in Figure C.19 – Figure C.26 accentuate the superior performance of the McOpt algorithms on various benchmarking test problems. Particularly, in the appearance

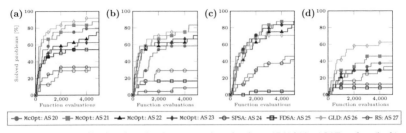

Figure 4.21. Data profiles for Algorithm Setup 20 – Algorithm Setup 27 (AS 20 – AS 27) as described in Appendix C.2 for the test functions and its noisy variations as described in Table C.1 with $\tau^* = 10^{-3}$. **Left (a)** Normal (J); **Middle left (b)** Multiplicative noise (J_{mult}); **Middle right (c)** Additive Gaussian noise (J_{add}); **Right (d)** Discontinuous (J_{dc}).

of additive Gaussian noise we observe a performance improvement w.r.t. the compared optimization algorithms. Also in terms of achieving a small normalized optimality gap, the presented algorithm class provides a very competitive approach (see Figure 4.20 (c)). Furthermore, we want to highlight Algorithm Setup 21 (see Appendix C.2), which belongs to characteristic **[C1]** and (3.40). In almost all of the considered 96 problems, this algorithm outperforms the other McOpt schemes. Note that for flat valleys, such as objectives like the *Rosenbrock* function, Algorithm Setup 21 (see Appendix C.2) has its weaknesses because of the scaled generating functions with the objective. As expected, the discontinuous variation of the test function is very challenging. Nevertheless, we can provide a competitive approach compared with the considered class of algorithms but must admit that there is space for improvement.

In summary, the data profiles and logarithmic optimality gap illustrations validate the presented McOpt algorithm class good performance and superior behavior with respect to noisy function evaluations.

4.6 Direct Policy Search for Reinforcement Learning

Before we conclude the numerical result chapter, we present an outlook and results on applying the McOpt algorithm class on reinforcement learning problems. In reinforcement learning a so-called policy for an agent in a given environment must be optimized such that a (unknown) reward function is maximized. From an optimization and control perspective, this leads to the task

$$\max_{u_t} \mathbb{E}_{e_t}\left\{ \sum_{t=0}^{N} J(z_t, u_t) \right\}$$

$$\text{s.t.} \quad z_{t+1} = a(z_t, u_t, e_t), \; z_0 = \bar{z} \tag{4.18}$$

where $z_t, \bar{z} \in \mathbb{R}^p$ is the system state, $u_t \in \mathbb{R}^q$ is the control input (and the decision variable), and $e_t \in \mathbb{R}^r$ is a random disturbance. Moreover, $a : \mathbb{R}^p \times \mathbb{R}^q \times \mathbb{R}^r \to \mathbb{R}^p$ is the state transition map and could be seen as the interacting environment in reinforcement learning terminology, where $J : \mathbb{R}^p \times \mathbb{R}^q \to \mathbb{R}$ is the reward function to be maximized. Well-known examples in this problem class are beating human standards in ATARI or even more complex video games and control locomotion tasks (see Mnih et al. (2013), Mnih et al. (2015), Todorov et al. (2012)). Conceptually, there exist three approaches to tackle (4.18). Namely, model-based reinforcement learning (system identification is incorporated with dynamic programming; see Polydoros and Nalpantidis (2017)), approximated dynamic programming (approximating the optimal control cost and apply it to dynamic programming; see Watkins and Dayan (1992)), and direct policy search (simulation/experiment based optimization; see Levine and Koltun (2013)). For a detailed and complete overview of reinforcement learning, we refer to Sutton and Barto (2018) and Henderson et al. (2018). Moreover, Recht (2019) shows the connection to control theory.

In recent years, derivative-free optimization algorithms have successfully been applied to reinforcement learning problems, specifically to the direct policy search approach as presented, for example, in Salimans et al. (2017), Lehman et al. (2018), and Mania et al. (2018). On the same lines, in Frihauf, Krstic, and Başar (2013) a discrete-time extremum seeking algorithm is applied to (4.18) (with $e_t = 0$) in its most simplest form, as also revisited by Dean, Mania, Matni, Recht, and Tu (2019) and Recht (2019). Specifically, (4.18) appears in the aforementioned work as a linear quadratic regulator problem given by

$$\max_{u_t} \mathbb{E}_{e_t} \left\{ \sum_{t=0}^{N-1} z_t^\top Q z_t + u_t^\top R u_t + z_N^\top S z_N \right\} \tag{4.19}$$
$$\text{s.t.} \quad z_{t+1} = A z_t + B u_t + e_t, \; z_0 = \bar{z}$$

with $Q, S, A \in \mathbb{R}^{p \times p}$, $R \in \mathbb{R}^{q \times q}$ and $B \in \mathbb{R}^{p \times q}$. Thus, linear system dynamics are disturbed by random noise and a quadratic reward function. Obviously, it is assumed that the reward function and the system dynamics are not known and that only measurements $J(z_t, u_t)$—term in the curly bracket in (4.19)—are available. This rather simple application must be seen more as a starting point; "though simple models are not the end of the story in analysis, it tends to be the case that if a complicated method fails to perform on a simple problem, then there is a flaw in the method", as pointed out in Recht (2019).

In the sequel we borrow (and adapt) the examples presented in Recht (2019) and Frihauf et al. (2013) where two different approaches to solve (4.19) are carried out. Note that instead of maximizing $J(z, u)$ we consider the opposite and minimizing the quadratic reward function. In the latter work, the elements of the open loop control sequence $\{u_k\}_{k=0}^{N-1}$ are the decision variables, whereas in Recht (2019) a constant and time-invariant control gain $K \in \mathbb{R}^{q \times p}$ has to be determined, such that $u_t = -K z_t$ can be applied. In both approaches a finite time (and open loop) control problem, i.e., $N < \infty$, is considered. Both examples are tackled with the following McOpt setup.

Algorithm Setup 9. McOpt with characteristic **[C1]** corresponding to (3.40) with

- Generating functions: $f(z) = \sin(z)$, $g(z) = \cos(z)$

- $T_d = \begin{bmatrix} Q & -I \\ I & Q \end{bmatrix} \in \mathbb{R}^{2n \times 2n}$ with Q as in Corollary 3, $n = pq$ or $n = N$, and singular values of W chosen to $\sigma = \{1, \ldots, 1\}$

and an adaptive step size ($h_0 = 0.01$, $\epsilon_1 = \epsilon_3 = 0.1$, $\epsilon_2 = 0.2$, $\nu_1 = 1.1$, $\nu_2 = 0.9$, and $\nu_3 = 0.8$) as described in Section 4.4.1 with randomized restarting of the exploration sequence ($m = 2n + 1$) as described in Section 4.4.2 is applied.   •

In the following, the two examples are compared with the cost given by solving the quadratic program (with the given matrices A, B, Q, R, S) or the optimal infinite control policy $u_t = -K^* x_t$, where K^* is calculated via the discrete algebraic Riccati equation (Zhou, Doyle, Glover, et al. (1996)), respectively. Additionally, the elements of the disturbance e_t are real-valued random variables sampled from the Gaussian distribution $\mathcal{N}(0, 10^{-4})$.

Finite LQ-Control—Frihauf et al. (2013)

A third-order unstable system is considered. The matrices in (4.19) are given by

$$A = \begin{bmatrix} 0.5 & 0.5 & 0.3 \\ 0.6 & 0.5 & -0.2 \\ 0.5 & 0 & 0.5 \end{bmatrix}, B = \begin{bmatrix} -0.15 \\ 0.5 \\ 0.75 \end{bmatrix},$$

$$Q_k = \gamma_k \begin{bmatrix} 1.0225 & 0.275 & 0 \\ 0.275 & 2.5 & 0 \\ 0 & 0 & 0 \end{bmatrix} \text{ with } \gamma_k = \{3, 5, 7, 9, 11, 1\}, S = Q_5, R = 5. \tag{4.20}$$

Note that the eigenvalues of the system matrix A are given by -0.22, 1.11, and 0.61. We set the control horizon to $N = 5$ and initialize the system with $z_0 = [-1\ 0.5\ 0.1]^\top$. The decision variables are set to $u_t = 0$ for $t = 0, \ldots, N - 1$. As one observes in Figure 4.22, the open loop control inputs $\{u_t\}_{t=0}^{N-1}$ converge in less than 200 iterations to their optimal values—almost a magnitude faster, as in Frihauf et al. (2013).

Data Center Cooling—Recht (2019)

In this problem we consider the matrices in (4.19) given by

$$A = \begin{bmatrix} 1.01 & 0.01 & 0 \\ 0.01 & 1.01 & 0.01 \\ 0 & 0.01 & 1.01 \end{bmatrix}, B = I, Q = I, S = I, R = 1000I. \tag{4.21}$$

It is introduced as a data center cooling problem by Gao (2014) in which three heat sources exist. The state z_t stands for the internal temperature of the heat sources affected by their neighbors. The sources can be controlled separately by a constant heat load. Note that two eigenvalues are slightly larger than 1; thus, the system is unstable for any initial

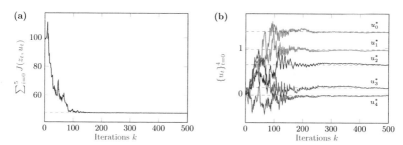

Figure 4.22. Application of the McOpt algorithm class to the finite LQ-control problem as in (4.19) (see Frihauf et al. (2013)) with unknown system dynamics and optimization objective given by the matrices in (4.20). **Left (a)** Objective value J; **Right (b)** Open loop control input sequence.

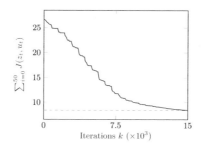

Figure 4.23. Objective value J for the application of the McOpt algorithm class to the data center cooling problem of the form as in (4.19) (see Gao (2014)) with unknown system dynamics and optimization objective given by the matrices in (4.21).

condition different to the zero vector. Therefore, approximating the system matrix A is very challenging because a small approximation error already leads to an unstable behavior in the controller design. We set the control horizon to $N = 50$ and initialize the system with $z_0 = [0.3\ 0.1\ 0.2]^\top$. The static control gain matrix is initialized with $K = 0 \in \mathbb{R}^{3\times3}$. As illustrated in Figure 4.23—median performance for 100 runs—McOpt converges to a cost function value as given by the optimal infinite control gain K^*.

As outlined in the beginning of this chapter, we kept the numerical experiments of the introduced application to direct policy search in reinforcement learning rather simple. However, the positive performance in these simple examples open the door to more complex and sophisticated problems in this class.

4.7 Conclusion and Outlook

In this chapter we provided the numerical results for the presented McOpt algorithm class. At first, we studied the various design parameters and functions on a qualitative numerical level as a counterpart to the theoretical investigations in Section 3.5. We gained an understanding how the behavior of the algorithm is affected by each parameter, interpreted the generating areas (as theoretical discussed in Section 3.3) and paved the way for tuning the algorithm performance. Beforehand, we ran a quantitative parameter study on the two-dimensional spherical function, where we analyzed the convergence speed and the averaged gradient approximation error. From an implementation and practical perspective, a faster convergence speed is sometimes gained with a bigger gradient approximation error. However, to ensure and accelerate convergence, we introduced variable and adaptive step size rules and a so-called (randomized) restarting of the exploration sequence. This led to performance improvements and the usage of larger step sizes, i.e., larger exploration of the objective, while ensuring convergence by restarting at the optimal visited point so far. Thereafter, we benchmarked four versions of the McOpt algorithm class against four other (local search) optimization schemes on eight different objective functions, plus their multiplicative and additive noise induced variations, as well as a discontinuous version—in total, 96 optimization problems. It turned out that McOpt is dominant on optimization problems where additive noise in the function evaluation is present. As a result of the explorative nature of the presented algorithm class, noise is averaged out—this underlines the observations in Wildhagen, Michalowsky, Feiling, and Ebenbauer (2018). Eventually, we provided an introduction how McOpt can be applied for reinforcement learning problems with direct policy search.

Outlook.   As introduced in Section 2.2, the presented optimization algorithm class is inspired by the concepts of extremum seeking control. Simultaneously, extremum seeking control provides a tremendous amount of potential applications (cf. Ariyur and Krstic (2003); Dochain, Perrier, and Guay (2011); Feiling, Koga, Krstić, and Oliveira (2018); Tan et al. (2010); Zhang and Ordóñez (2011), to name only a few). As presented in Feiling et al. (2019), only the single-point characteristic is eligible for application to the real-time optimization problem class of extremum seeking control. Hence, a further rather theoretical extension and research direction is the development of the multi-variable case—aside from a coordinate descent—of the two-point characteristic presented in Feiling et al. (2019). However, the aforementioned practical applications can be tackled by the presented algorithm class McOpt. Implementing our approach might be even more efficient and simpler, since real-world applications are given in discrete-time, whereas most extremum seeking schemes are formulated in continuous-time. Obviously, there are many more fields of applications such as those introduced in Section 2.1. If noisy measurements or local minima are present in the considered optimization problem, the McOpt algorithm class might be a fruitful choice.

5

Summary and Conclusion

In this thesis, we developed a novel procedure to extract gradient information of an objective function. In particular, we constructed non-commutative maps based on gradient-generating functions and periodic exploration sequences. The compositions of those maps lead to approximations of gradient descent steps; thereby, we derived the presented class of derivative-free optimization algorithms, which we entitled McOpt. The *Rukik's Cube* is the eponym of the algorithm class, with its underling mathematical concept of non-commutativity, as illustrated in Figure 5.1. The main features of this algorithm class, as validated by our numerical benchmarking study, are its robustness against noisy function evaluations and its ability to deal with discontinuous objective functions. Moreover, its exploratory behavior potentially leads to overcoming local minima.

In Chapter 2 we posed the main idea of the McOpt algorithm class, which is motivated by the continuous-time control concept extremum seeking. In particular, one approach is based on approximating Lie brackets of vector fields. As known from non-linear geometric control, Lie brackets are a measurement of the non-commutativity of flows. Choosing the vector fields with an objective function as their argument in a proper way, the non-commutativity gap, in the form of Lie brackets, leads to the negative gradient of the objective function. We utilized this concept and developed a similar procedure for the discrete-time setting. Specifically, a method to approximate the flow maps was presented and combined with the introduced periodic exploration sequence. Then we showed in a basic example by applying Taylor's Theorem that the composition of the derived non-commutative maps yields to the gradient approximation of the objective function; thereby, to a gradient descent step as depicted in Figure 5.1.

In Chapter 3 we studied the theoretical problem of the presented algorithm class. In essence, we derived a generalized and parameterized version of the McOpt algorithm class and investigated its gradient descent approximation behavior. This led us to a set of nonlinear equations which were key to designing the algorithm's ingredients: periodic exploration sequences and pairs of gradient-generating functions. First, periodic exploration sequences were obtained by solving a system of quadratic matrix equations—which we encountered by a singular value decomposition—under certain structural matrix conditions, such as normality, definiteness, skew-symmetry, and interlacing eigenvalue properties. Second, several classes of gradient-generating functions were derived by solving a set of functional equations. Based on the parametrization and the structural matrix conditions,

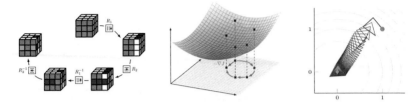

Figure 5.1. The *Rubik's Cube* stands as figurative illustration for non-commutative maps and as eponym of the presented McOpt (**M**agic **C**ube **Opt**imization) algorithm class. Non-commutative maps are constructed such that their compositions reveal an approximation of the gradient of an objective function. This leads to an explorative behavior and convergence to the neighborhood of a local minimum.

two characteristics of the McOpt algorithms were proposed, so-called single-point and two-point characteristics, implying that the objective must be evaluated once or twice per iteration. Processing one more function evaluation per iteration offers more degrees of freedom in choosing gradient-generating functions and the introduced design parameters. The last part of this chapter dealt specifically with the design parameters and how they influence the exploration sequence. Moreover, we proved semi-global practical asymptotic convergence for the presented algorithm class with a fixed step size and we proved semi-global asymptotic convergence by introducing a decreasing step size.

Chapter 4 was dedicated to numerical results of the presented algorithm class. We provided a qualitative and quantitative study of the design parameters and functions. We focused on the influence of these parameters on the algorithm's exploration and convergence behavior as depicted in Figure 5.1. We carried out a case-by-case investigation for each parameter. Moreover, we discussed well-performing settings of parameters based on convergence speed and gradient approximation error. Based on these insights, we proposed a decreasing and adaptive step size rule such that asymptotic convergence rather than practical convergence is achieved while convergence speed is accelerated by larger (initial) step sizes. An adaption and randomization of the exploration sequence led to further performance improvements. Eventually, various numerical experiments were presented. First, we presented a benchmarking study of the McOpt algorithm class on several challenging multi-variable, non-convex, ill-conditioned, and discontinuous optimization problems against other derivative-free optimization algorithms. Second, we presented one of many potential applications of McOpt; namely, how it can be utilized for direct policy search in reinforcement learning problems. The experimental and numerical results underlined the good performance in the presence of noisy function evaluations.

Future research could take several different directions. From a theoretical perspective, extensions of the constructive exploration sequence approach to non-holonomic control problems, inverse signature path problems, or higher-level signature path constructions could be explored as well as application of the presented gradient approximation procedure to gradient descent algorithms with momentum as discussed in Section 3.3 and Section 3.6. From a numerical and application point of view, we have already discussed different angles in Section 4.7.

A

Notation, Background, and Preliminary Lemmas

A.1 Notation

In the following we introduce the notation of this thesis. A list of the nomenclature is provided on page 123 including specific notation of the presented algorithm class McOpt.

Sets. We denote by $\mathbb{R}$, $\mathbb{R}_{\geq 0}$, and $\mathbb{R}_{>0}$ the set of reals, non-negative reals, and positive reals, respectively. $\mathbb{R}^n$ stands for the n-dimensional vector space. By $\mathbb{N}$ we denote the natural numbers excluding zero and define $\mathbb{N}_0 := \mathbb{N} \cup \{0\}$. Further, $\mathbb{C}$ denotes the set of complex numbers. A closed interval is denoted by $[a, b] := \{x \in \mathbb{R} : a \leq x \leq b\}$. Similar, we define the semi-closed interval as $[a, b[:= \{x \in \mathbb{R} : a \leq x < b\}$, and the open interval as $]a, b[:= \{x \in \mathbb{R} : a < x < b\}$. A compact set with center point $x^* \in \mathbb{R}^n$ and radius $\delta \in \mathbb{R}_{\geq 0}$ is denoted by $\mathcal{U}_\delta(x^*) \subseteq \mathbb{R}^n$ and defined as $\{x \in \mathbb{R}^n : \|x - x^*\|_2 \leq \delta\}$.

Functions. The set of k-times differentiable functions is denoted by $C^k(\mathbb{R}^n, \mathbb{R})$. Further, we denote the gradient of $J : \mathbb{R}^n \to \mathbb{R}$ with $\nabla J(x) := [\partial J/\partial x_1 \ \cdots \ \partial J/\partial x_n]$.

Vectors and Matrices. We denote the i-th unit vector by e_i; all entries are zero, except the i-th entry is one. The unit matrix is denoted by $I \in \mathbb{R}^{n \times n}$, i.e., a matrix with diagonal elements equal to one and the rest are equal to zero. An all-one vector, where all entries are one, is given by $\mathbb{1}$. Moreover, $(P)_{1:r} \in \mathbb{R}^{m \times n}$ stands for the $r \times r$ (leading) principle submatrix of P, i.e., the first r rows and columns of P with $r \leq \min\{m, n\}$. The rank of a matrix P is denoted by $\mathrm{rk}(P)$.

Operators and Sequences. A composition between two functions $f, g : \mathbb{R}^n \to \mathbb{R}$ is defined as $x \mapsto (f \circ g)(x) := f(g(x))$. Let $f(x; h) = \mathcal{O}(h)$, i.e., for every compact set $\mathcal{X} \subseteq \mathbb{R}^n$ there exist a $L \in \mathbb{R}_{>0}$ and $\bar{h} \in \mathbb{R}_{>0}$, such that for all $x \in \mathcal{X}$ and $h \in [0, \bar{h}]$, $\|f(x; h)\|_2 \leq Lh$. We denote the Lie bracket of two continuously differentiable vector fields $f, g : \mathbb{R}^n \to \mathbb{R}^n$ by $[f, g] : z \mapsto \partial g/z(z) f(z) - \partial f/z(z) g(z)$. The ceiling and floor operator are defined as $\lfloor x \rfloor := \max\{k \in \mathbb{Z} \mid k \leq x\}$ and $\lceil x \rceil := \min\{k \in \mathbb{Z} \mid k \geq x\}$, respectively.

The operator mod takes to integers k and n and returns an integer $k \bmod n$, equal to the remainder of the division of k by n. Further, we denote a sequence of length $m + 1$ by $\{x_k\}_{k=0}^{m} := x_0, x_1, \ldots, x_m$.

A.2 Technical Background

In this section we recall several technical concepts that are utilized in this thesis.

Lie Brackets

In the following we provide an introduction to Lie brackets. For a more detailed presentation of this topic we refer to Bourbaki (2008) or other textbooks in this field. We emphasize that Lie brackets are defined as an operator on vector fields. In this work the vector fields (i.e., $f, g : \mathbb{R}^n \to \mathbb{R}^n$) are replaced by scalar functions (i.e., $f, g : \mathbb{R} \to \mathbb{R}$). Nevertheless, we give the introduction for the general case (cf. Definition 1):

Let $f, g, h : \mathbb{R}^n \to \mathbb{R}^n$ be continuously differentiable vector fields. Then we define

$$[f, g] : z \to \frac{\partial g}{\partial z}(z) f(z) - \frac{\partial g}{\partial z}(z) f(z) \tag{A.1}$$

as the Lie bracket of f, g. Lie brackets have the following properties:

1. Bilinearity: For all $a, b \in \mathbb{R}$ we have for all $z \in \mathbb{R}^n$

$$[af + bg, h](z) = a[f, h](z) + b[g, h](z). \tag{A.2}$$

2. Skew-symmetry: We have for all $z \in \mathbb{R}^n$

$$[f, g](z) = -[g, f](z). \tag{A.3}$$

3. Jacobi-identity: We have for all $z \in \mathbb{R}^n$

$$[f, [g, h]](z) = -[g, [h, f]](z) - [h, [f, g]](z). \tag{A.4}$$

Special Taylor Expansion

The following lemma states a second order (general) Taylor expansion with specific argument of the function to be expanded. It is utilized for the results in Theorem 1 in Section 3.1 and Lemma 6 in Appendix A.3.

Lemma 3. Let $a \in C^2(\mathbb{R}^n; \mathbb{R})$, $b \in C^0(\mathbb{R}^p; \mathbb{R}^n)$, and $h \in \mathbb{R}_{\geq 0}$. Then for any compact convex set $\mathcal{Z} \subseteq \mathbb{R}^n$ and any compact set $\mathcal{Y} \subseteq \mathbb{R}^p$ there exist a $R \in C^0(\mathbb{R}^n \times \mathbb{R}^p \times \mathbb{R}_{\geq 0}; \mathbb{R})$ and a $M \in \mathbb{R}_{\geq 0}$ such that for all z, $z + hb(y) \in \mathcal{Z}$ and $y \in \mathcal{Y}$ we have

$$a(z + hb(y)) = a(z) + h\frac{\partial a}{\partial z}(z)^\top b(y) + R(z, y; h^2) \tag{A.5}$$

with $|R(z, y; h^2)| \leq Mh^2$, i.e., $\lim_{h \to 0} R(z, y; h^2) = 0$. $\qquad\bullet$

Proof. Equation (A.5) is obtained by applying Taylor's theorem (Rudin, 1964, Theorem 5.15, p. 110) up to degree two, thus, there exits a $\theta \in [0,1]$ such that

$$a(z + hb(y)) = a(z) + h\frac{\partial a}{\partial z}(z)^\top b(y) + \frac{h^2}{2}b(y)^\top \frac{\partial^2 a}{\partial z}(\bar{x})b(y) \tag{A.6}$$

holds with $\bar{x} = z + \theta hb(y)$. The term

$$R(z, y; h^2) = h^2 b(y)^\top \frac{\partial^2 a}{\partial^2 z}(\bar{x})b(y) \tag{A.7}$$

is the Lagrange remainder where $\partial^2 a/\partial^2 z(\cdot)$ is the Hessian of $a(\cdot)$. Since $\mathcal{Z}$ and $\mathcal{Y}$ are compact and $b \in C^0(\mathbb{R}^p; \mathbb{R}^n)$ there exists a $M_b \in \mathbb{R}_{\geq 0}$ such that

$$\|b(y)\|_2 \leq M_b, \quad y \in \mathcal{Y} \tag{A.8}$$

holds. Furthermore, since $\mathcal{Z}$ is convex and compact and $a \in C^2(\mathbb{R}^n; \mathbb{R})$ there exists a $M_a \in \mathbb{R}_{\geq 0}$ such that

$$\|\frac{\partial^2 a}{\partial z^2}(\bar{x})\|_2 \leq M_a, \quad \bar{x} \in \mathcal{Z}. \tag{A.9}$$

Finally, $|R(z, y; h^2)| \leq Mh^2$ with $M = M_b^2 M_a$. □

Matrix Properties

In this section we recall some basic matrix properties of normal and skew-symmetric matrices. We refer the reader to Horn and Johnson (2012) or other text books on linear algebra and matrix theory for a more detailed introduction.

Normal Matrices. Let $A \in \mathbb{R}^{n \times n}$ be a normal matrix, then

$$AA^\top = A^\top A \tag{A.10}$$

holds. Any real normal matrix has conjugate complex eigenvalue pairs $\gamma_\ell \pm \delta_\ell i$ with $\gamma_\ell \in \mathbb{R}$ and $\delta_\ell \in \mathbb{R}_{\geq 0}$ for $\ell = 1, \ldots, \lceil n/2 \rceil$. Moreover, every normal matrix is diagonalizable by an unitary matrix $U \in \mathbb{R}^{n \times n}$ (with $UU^\top = U^\top U = I$) such that the spectral orthogonal transformation

$$UAU^\top = \mathrm{diag}([D_1\, D_2\, \cdots\, D_{\lceil n/2 \rceil}]) \quad \text{with} \quad D_i = \begin{bmatrix} \gamma_i & -\delta_i \\ \delta_i & \gamma_i \end{bmatrix} \tag{A.11}$$

is given. Note that $D_{\lceil n/2 \rceil} = \gamma_{\lceil n/2 \rceil}$ if n odd. The unitary matrix U can be constructed by the eigenvectors $a_\ell \pm b_\ell i$ with $a_\ell, b_\ell \in \mathbb{R}^n$ of A as

$$\begin{aligned} U &= \begin{bmatrix} a_1 & b_1 & a_2 & b_2 & \cdots & a_{n/2} & b_{n/2} \end{bmatrix} \quad \text{for } n \text{ even,} \\ U &= \begin{bmatrix} a_1 & b_1 & a_2 & b_2 & \cdots & a_{\lceil n/2 \rceil} \end{bmatrix} \quad \text{for } n \text{ odd.} \end{aligned} \tag{A.12}$$

Skew-Symmetric Matrices. Let $A \in \mathbb{R}^{n \times n}$ be a skew-symmetric matrix, then

$$A^\top = -A \qquad (A.13)$$

holds. Thus, skew-symmetric matrices are a special case of normal matrices. Any real skew-symmetric matrix has conjugate complex eigenvalue pairs $\pm \delta_\ell i$ with $\delta_\ell \in \mathbb{R}_{\geq 0}$ for $\ell = 1, \ldots, \lceil n/2 \rceil$. Therefore, the spectral orthogonal transformation (A.11) with (A.12) holds (where $\gamma_\ell = 0$).

Eigenvalue Interlacing Properties. The following two lemmas state the sufficient part of Cauchy's (eigenvalue) interlacing inequalities (cf. Cauchy (1891)) for real skew-symmetric matrices. Thus, the eigenvalues of the principal submatrix can be chosen w.r.t. certain inequalities depending on the eigenvalues of the given skew-symmetric matrix.

Lemma 4. Let $C \in \mathbb{R}^{p \times p}$ be a skew-symmetric matrix with eigenvalues $\pm \eta_k i, \eta_k \in \mathbb{R}_{\geq 0}, k = 1, \ldots, \lceil p/2 \rceil$, and let $\omega_\ell \in \mathbb{R}_{\geq 0}, \ell = 1, \ldots, \lceil p/2 \rceil - 1$ be such that the inequalities

$$\eta_1 \geq \omega_1 \geq \eta_2 \geq \omega_2 \cdots \geq \eta_{\lceil p/2 \rceil - 1} \geq \omega_{\lceil p/2 \rceil - 1} \geq \eta_{\lceil p/2 \rceil} \geq 0, \qquad (A.14)$$

are satisfied. Then there exists an unitary matrix $\Theta \in \mathbb{R}^{p \times p}$ such that $Q \in \mathbb{R}^{(p-1) \times (p-1)}$ is a principal submatrix of $\Theta^\top C \Theta$ with eigenvalues $\pm \omega_\ell i$. •

For a proof, we refer to Thompson (1979).

Lemma 5. Let $C \in \mathbb{R}^{p \times p}$ be a skew-symmetric matrix with eigenvalues $\pm \eta_\ell i, \eta_\ell \in \mathbb{R}_{\geq 0}, \ell = 1, \ldots, p$ arranged according to

$$\eta_1 \geq \eta_2 \geq \ldots \geq \eta_{\lceil p/2 \rceil} \geq 0. \qquad (A.15)$$

Then for $\omega_1 \geq \omega_2 \geq \ldots \geq \omega_r$ with $\omega_k \in \mathbb{R}_{\geq 0}$ such that

$$\eta_k \geq \omega_k \geq \eta_{\lceil p/2 \rceil - r + k}, \qquad (A.16)$$

there exists a unitary matrix $\Theta \in \mathbb{R}^{p \times p}$ such that $Q \in \mathbb{R}^{(2r) \times (2r)}$ is a principal submatrix of $\Theta^\top C \Theta$ with eigenvalues $\pm \omega_\ell i, \ell = 1, \ldots, r$. •

Proof. Applying Lemma 4 $\lceil p/2 \rceil - r$ times yield the result, similar to the proof of (Fan & Pall, 1957, Theorem 1). □

A.3 Preliminary Lemmas and Results

Lemma 6. Let Assumption [A1] hold true. Moreover, let $\mathcal{X} \subseteq \mathbb{R}^n$ and $\mathcal{J} \subseteq \mathbb{R}$ be compact convex sets, and $m \in \mathbb{N}_{\geq 1}$. Then there exist a function $R_{k+m-1} : \mathbb{R}^n \times \mathbb{R} \times \mathbb{R}_{\geq 0} \to \mathbb{R}^n$ and a constant $M_{k+m-1} \in \mathbb{R}_{\geq 0}$ such that for any iterates $x_k, \ldots, x_{k+m}$ of the algorithm

(3.1) with maps in (3.2), $x_t, x_t + \sqrt{h}s_k(J(x_t)) \in \mathcal{X}$, and $J(x_t), J(x_t + \sqrt{h}s_k(J(x_t))) \in \mathcal{J}$ for $t = k, \ldots, k + m$, and we have

$$x_{k+m} = x_k + \sqrt{h}(\alpha_1 + \alpha_2) \sum_{i=k}^{k+m-1} s_i(J(x_k)) + h\alpha_2 \sum_{i=k}^{k+m-1} \frac{\partial s_i}{\partial J}(J(x_k))s_i(J(x_k))^\top \nabla J(x_k)$$

$$+ h(\alpha_1 + \alpha_2)^2 \sum_{i=k}^{k+m-1} \sum_{j=k}^{i-1} \frac{\partial s_i}{\partial J}(J(x_k))s_j(J(x_k))^\top \nabla J(x_k) + R_{m-1}(x_k, J(x_k); h^{3/2}), \quad \text{(A.17)}$$

with $\|R_{k+m-1}(x_k, J(x_k); h^{3/2})\|_2 \le M_{k+m-1}h^{3/2}$, i.e., $R_{k+m-1}(x_k, J(x_k); h^{3/2}) = \mathcal{O}(h^{3/2})$. $\quad \bullet$

Proof. W.l.o.g we set $k = 0$, i.e., we show by induction that the m-step evolution of (3.1) with transition map (3.2) is give by (A.17) with $k = 0$. Similarly to $R(\cdot, \cdot; \cdot)$ in (A.5), we introduce the following notation of the *Taylor remainder (T.R.)* terms for $k = 0, \ldots, m - 1$:

- $R_{J,k}(\cdot, \cdot; h)$ of T.R. of $J(x_k)$

- $R_{s,k}(\cdot, \cdot; h)$ of T.R. of $s_k(J(x_k))$

- $R_{J,k}^+(\cdot, \cdot; h)$ of T.R. of $J(x_k + \sqrt{h}s_k(J(x_k)))$

- $R_{s,k}^+(\cdot, \cdot; h)$ of T.R. of $s_k(J(x_k + \sqrt{h}s_k(J(x_k))))$

and aggregated remainders with terms of order h or $h^{3/2}$ and higher for $k = 0, \ldots, m - 1$:

- $R_k(\cdot, \cdot; h^{3/2})$ of x_k

- $\bar{R}_{J,k}(\cdot, \cdot; h)$ of $J(x_k)$

- $\bar{R}_{s,k}(\cdot, \cdot; h)$ of $s_k(J(x_k))$

- $\bar{R}_{J,k}^+(\cdot, \cdot; h)$ of $J(x_k + \sqrt{h}s_k(J(x_k)))$

- $\bar{R}_{s,k}^+(\cdot, \cdot; h)$ of $s_k(J(x_k + \sqrt{h}s_k(J(x_k))))$

Step 1: Basis. Consider the first step of (3.1) with (3.2), i.e,

$$x_1 = x_0 + \sqrt{h}\alpha_1 s_0(J(x_0)) + \sqrt{h}\alpha_2 s_0(J(x_0 + \sqrt{h}s_0(J(x_0))))$$

$$\overset{\text{(Lemma 3)}}{=} x_0 + \sqrt{h}(\alpha_1 + \alpha_2)s_0(J(x_0)) + h\alpha_2 \frac{\partial s_0}{\partial J}(J(x_0))s_0(J(x_0))^\top \nabla J(x_0)$$

$$+ R_0(x_0, J(x_0); h^{3/2}) \tag{A.18}$$

with $R_0 : \mathbb{R}^n \times \mathbb{R} \times \mathbb{R}_{\ge 0} \to \mathbb{R}^n$. In the above equation, Lemma 3 is applied twice. First, for $J(x_0 + \sqrt{h}s_0(J(x_0)))$ where $a(\cdot) := J(\cdot), b(\cdot) := s_0(\cdot), z := x_0$, and $y := J(x_0)$ in Lemma 3 are chosen such that $R_{J,0}^+(\cdot, \cdot; h)$ as in (A.7) exists, i.e.,

$$J(x_0 + \sqrt{h}s_0(J(x_0))) = J(x_0) + \sqrt{h}s_0(J(x_0))^\top \nabla J(x_0) + R_{J,0}^+(x_0, J(x_0); h) \tag{A.19}$$

with $R_{J,0}^+ : \mathbb{R}^n \times \mathbb{R} \times \mathbb{R}_{\geq 0} \to \mathbb{R}$. Second, for $s_0(J(x_0 + \sqrt{h}s_0(J(x_0))))$ where $a(\cdot) := f(\cdot)$ and $a(\cdot) := g(\cdot)$, $b(x_0, J(x_0)) = s_0(J(x_0))^\top \nabla J(x_0) + R_{J,0}^+(x_0, J(x_0); h)$, $z := J(x_0)$, and $y = [x_0^\top \, J(x_0)]^\top$ in Lemma 3 are chosen such that a $R_{s,0}^+(\cdot, \cdot; h)$ as in (A.7) exists with given bounds on u_0 and v_0, i.e.,

$$s_0\big(J(x_0 + \sqrt{h}s_0(J(x_0)))\big) = s_0\Big(J(x_0) + \sqrt{h}s_0(J(x_0))^\top \nabla J(x_0) + R_{J,0}^+(x_0, J(x_0); h)\Big)$$

$$= s_0(J(x_0)) + \frac{\partial s_0}{\partial J}(J(x_0))\Big(\sqrt{h}s_0(J(x_0))^\top \nabla J(x_0) + R_{J,0}^+(x_0, J(x_0); h)\Big)$$

$$+ R_{s,0}^+(x_0, J(x_0); h) \tag{A.20}$$

with $R_{s,0}^+ : \mathbb{R}^n \times \mathbb{R} \times \mathbb{R}_{>0} \to \mathbb{R}^n$. Note that $s_0(\cdot) = f(\cdot)u_0 + g(\cdot)v_0$ (see (3.2)), i.e., (A.20) is obtained by applying Lemma 3 to $f(\cdot)$ and $g(\cdot)$ separately. We neglected this intermediate step and directly stated the Taylor expansion for $s_0(\cdot)$. In the sequel we do not highlight this intermediate step and "directly" apply Lemma 3 to $a(\cdot) := s_i(\cdot)$ which implies that Lemma 3 is applied to $a(\cdot) := f_i(\cdot)$ and $a(\cdot) := g_i(\cdot)$ separately with given bounds on u_i and v_i.

Then, since $J \in C^2(\mathbb{R}^n; \mathbb{R})$ by Assumption **[A1]**, $x_0, x_0 + \sqrt{h}s_0(J(x_0)) \in \mathcal{X}$, and $J(x_0) \in \mathcal{J}$, we conclude with Lemma 3, there exists a $M_{J,0}^+ \in \mathbb{R}_{\geq 0}$ such that $\|R_{J,0}^+(x_0, J(x_0); h)\|_2 \leq M_{J,0}^+h$. Additionally, by assumption $s_0 \in C^2(\mathbb{R}; \mathbb{R}^n)$ (see Assumption **[A1]**) and $J(x_0 + \sqrt{h}s_0(J(x_0))) \in \mathcal{J}$, thus, by Lemma 3 there exists a $M_{s,0}^+ \in \mathbb{R}_{\geq 0}$ such that $\|R_{s,0}^+(x_0, J(x_0); h)\|_2 \leq M_{s,0}^+h$. Putting these facts together, we obtain that

$$R_0(x_0, J(x_0); h^{3/2}) = \sqrt{h}\frac{\partial s_0}{\partial J}(J(x_0)R_{J,0}^+(x_0, J(x_0); h) + \sqrt{h}R_{s,0}^+(x_0, J(x_0); h) \tag{A.21}$$

in (A.18) holds. Then by assumption $s_0 \in C^2(\mathbb{R}; \mathbb{R}^n)$ (see Assumption **[A1]**) and $J(x_0) \in \mathcal{J}$, there exits a $L_{s,0} \in \mathbb{R}_{\geq 0}$ such that

$$\|\frac{\partial s_0}{\partial J}(J(x_0)\|_2 \leq L_{s,0}. \tag{A.22}$$

Then it follows that

$$\|R_0(x_0, J(x_0); h^{3/2})\|_2 \leq M_0h^{3/2} := (M_{J,0}^+L_{s,0} + M_{s,0}^+)h^{3/2} \tag{A.23}$$

and we can obtain that $R_0(x_0, J(x_0); h^{3/2}) = \mathcal{O}(h^{3/2})$, thus, (A.18) is (A.17) for $k = 0$ and $m = 1$.

   Step 2: Inductive Step. Assume that (A.17) holds for x_{m-1}, i.e., that the evolution of x_k for $k = 0, \dots, m-1$ reads

$$x_{m-1} = x_0 + \sqrt{h}(\alpha_1 + \alpha_2)\sum_{i=0}^{m-2} s_i(J(x_0)) + h\alpha_2 \sum_{i=0}^{m-2} \frac{\partial s_i}{\partial J}(J(x_0))s_i(J(x_0))^\top \nabla J(x_0)$$

$$+ h(\alpha_1 + \alpha_2)^2 \sum_{i=0}^{m-2}\sum_{j=0}^{i-1} \frac{\partial s_i}{\partial J}(J(x_0))s_j(J(x_0))^\top \nabla J(x_0) + R_{m-2}(x_0, J(x_0); h^{3/2}) \tag{A.24}$$

and there exists a $M_{m-2} \in \mathbb{R}_{\geq 0}$ such that $\|R_{m-2}(x_0, J(x_0); h^{3/2})\|_2 \leq M_{m-2}h^{3/2}$. Next we consider the m-th step of (3.1) with (3.2), i.e.,

$$x_m = x_{m-1} + \sqrt{h}\alpha_1 s_{m-1}(J(x_{m-1})) + \sqrt{h}\alpha_2 s_{m-1}\Big(J\big(x_{m-1} + \sqrt{h}s_{m-1}(J(x_{m-1}))\big)\Big). \quad (A.25)$$

Again, as in Step 1, we apply Lemma 3 several times. First for $J(x_{m-1})$ where $a(\cdot) := J(\cdot)$, $b(x_0, J(x_0)) = h^{-1/2}(rhs.\ of\ (A.24) - x_0)$, $z := x_0$, and $y = [x_0^\top J(x_0)]^\top$ in Lemma 3 are chosen such that a $R_{J,m-1}(\cdot, \cdot; h)$ as in (A.7) exists, i.e.,

$$J(x_{m-1}) = J(r.h.s.\ of\ (A.24))$$

$$\overset{\text{(Lemma 3)}}{=} J(x_0) + \sqrt{h}(\alpha_1 + \alpha_2)\sum_{i=0}^{m-2} s_i(J(x_0))^\top \nabla J(x_0) + \bar{R}_{J,m-1}(x_0, J(x_0); h) \quad (A.26)$$

with $\bar{R}_{J,m-1} : \mathbb{R}^n \times \mathbb{R} \times \mathbb{R}_{>0} \to \mathbb{R}$ where

$$\bar{R}_{J,m-1}(x_0, J(x_0); h) = h\nabla J(x_0)^\top \Big(\alpha_2 \sum_{i=0}^{m-2} s_i(J(x_0))\frac{\partial s_i}{\partial J}(J(x_0))^\top$$

$$+ (\alpha_1 + \alpha_2)^2 \sum_{i=0}^{m-2}\sum_{j=0}^{i-1} s_j(J(x_0))\frac{\partial s_i}{\partial J}(J(x_0))^\top\Big)\nabla J(x_0)$$

$$+ R_{m-2}(x_0, J(x_0); h^{3/2}) + R_{J,m-1}(x_0, J(x_0); h) \quad (A.27)$$

with $R_{J,m-1} : \mathbb{R}^n \times \mathbb{R} \times \mathbb{R}_{\geq 0} \to \mathbb{R}$. Secondly, Lemma 3 is applied on $s_{m-1}(J(x_{m-1}))$ where $a(\cdot) := s_{m-1}(\cdot), b(x_0, J(x_0)) = h^{-1/2}(rhs.\ of\ (A.26) - J(x_0)), z := J(x_0)$, and $y = [x_0^\top J(x_0)]^\top$ in Lemma 3 are chosen such that a $R_{s,m-1}(\cdot, \cdot; h)$ as in (A.7) exists, i.e.,

$$s_{m-1}(J(x_{m-1})) = s_{m-1}(rhs.\ of\ (A.26))$$

$$\overset{\text{(Lemma 3)}}{=} s_{m-1}(J(x_0)) + \sqrt{h}(\alpha_1 + \alpha_2)\sum_{i=0}^{m-2} \frac{\partial s_{m-1}}{\partial J}(J(x_0))s_i(J(x_0))^\top \nabla J(x_0)$$

$$+ \bar{R}_{s,m-1}(x_0, J(x_0); h) \quad (A.28)$$

with $\bar{R}_{s,m-1} : \mathbb{R}^n \times \mathbb{R} \times \mathbb{R}_{\geq 0} \to \mathbb{R}^n$ where

$$\bar{R}_{s,m-1}(x_0, J(x_0); h) = \frac{\partial s_{m-1}}{\partial J}(J(x_0)\bar{R}_{J,m-1}(x_0, J(x_0); h) + R_{s,m-1}(x_0, J(x_0); h) \quad (A.29)$$

with $R_{s,m-1} : \mathbb{R}^n \times \mathbb{R} \times \mathbb{R}_{\geq 0} \to \mathbb{R}^n$. Thirdly, Lemma 3 is applied on $J(x_{m-1} + \sqrt{h}s_{m-1}(J(x_{m-1})))$ where $a(\cdot) := J(\cdot), b(x_0, J(x_0)) = h^{-1/2}(rhs.\ of\ (A.24) - x_0) + \sqrt{h}(rhs.\ of\ (A.28)), z := x_0$, and $y = [x_0^\top J(x_0)]^\top$ in Lemma 3 are chosen such that a $R_{J,m-1}^+(\cdot, \cdot; h)$ as in (A.7) exists, i.e.,

$$J(x_{m-1} + \sqrt{h}s_{m-1}(J(x_{m-1}))) = J(rhs.\ of\ (A.24) + \sqrt{h}(rhs.\ of\ (A.28)))$$

$$\overset{\text{(Lemma 3)}}{=} J(x_0) + \sqrt{h}(\alpha_1 + \alpha_2)\sum_{i=0}^{m-2} s_i(J(x_0))^\top\nabla J(x_0) + \sqrt{h}s_{m-1}(J(x_0))^\top\nabla J(x_0)$$

$$+ \bar{R}_{J,m-1}^+(x_0, J(x_0); h) \quad (A.30)$$

with $\bar{R}^+_{J,m-1} : \mathbb{R}^n \times \mathbb{R} \times \mathbb{R}_{\geq 0} \to \mathbb{R}$ where

$$\bar{R}^+_{J,m-1}(x_0, J(x_0); h) = h\nabla J(x_0)^\top \Big(\alpha_2 \sum_{i=0}^{m-2} s_i(J(x_0)) \frac{\partial s_i}{\partial J}(J(x_0))^\top$$

$$+ (\alpha_1 + \alpha_2)^2 \sum_{i=0}^{m-2}\sum_{j=0}^{i-1} s_j(J(x_0)) \frac{\partial s_i}{\partial J}(J(x_0))^\top (\alpha_1 + \alpha_2) \sum_{i=0}^{m-2} s_i(J(x_0)) \frac{\partial s_{m-1}}{\partial J}(J(x_0))^\top \Big) \nabla J(x_0)$$

$$+ R_{m-2}(x_0, J(x_0); h^{3/2}) + \sqrt{h}\bar{R}_{s,m-1}(x_0, J(x_0); h) + R^+_{J,m-1}(x_0, J(x_0); h) \tag{A.31}$$

with $R^+_{J,m-1} : \mathbb{R}^n \times \mathbb{R} \times \mathbb{R}_{\geq 0} \to \mathbb{R}$. Lastly, Lemma 3 is applied on $s_{m-1}(J(x_{m-1} + \sqrt{h}s_{m-1}(J(x_{m-1}))))$ where $a(\cdot) := s_{m-1}(\cdot)$, $b(x_0, J(x_0)) = h^{-1/2}(rhs.\ of\ (A.30) - J(x_0))$, $z := J(x_0)$, and $y = [x_0^\top\ J(x_0)]^\top$ in Lemma 3 are chosen such that a $R^+_{s,m-1}(\cdot,\cdot;h)$ as in (A.7) exists, i.e.,

$$s_{m-1}(J(x_{m-1} + \sqrt{h}s_{m-1}(J(x_{m-1})))) = s_{m-1}(rhs.\ of\ (A.30))$$

$$\overset{(\text{Lemma 3})}{=} s_{m-1}(J(x_0)) + \sqrt{h}(\alpha_1 + \alpha_2) \sum_{i=0}^{m-2} \frac{\partial s_{m-1}}{\partial J}(J(x_0))s_i(J(x_0))^\top \nabla J(x_0)$$

$$+ \sqrt{h}\frac{\partial s_{m-1}}{\partial J}(J(x_0))s_{m-1}(J(x_0))^\top \nabla J(x_0) + \bar{R}^+_{s,m-1}(x_0, J(x_0); h) \tag{A.32}$$

with $\bar{R}^+_{s,m-1} : \mathbb{R}^n \times \mathbb{R} \times \mathbb{R}_{>0} \to \mathbb{R}^n$ where

$$\bar{R}^+_{s,m-1}(x_0, J(x_0); h) = \frac{\partial s_{m-1}}{\partial J}(J(x_0)\bar{R}^+_{J,m-1}(x_0, J(x_0); h) + R^+_{s,m-1}(x_0, J(x_0); h) \tag{A.33}$$

with $R^+_{s,m-1} : \mathbb{R}^n \times \mathbb{R} \times \mathbb{R}_{\geq 0} \to \mathbb{R}^n$.

Then with the same arguments as in Step 1, namely by assumption $J \in C^2(\mathbb{R}^n; \mathbb{R})$ (see Assumption [A1]) for every $k = 0, \ldots, m-1$, $s_k \in C^2(\mathbb{R}; \mathbb{R}^n)$ (see Assumption [A1]), x_k, $x_k + \sqrt{h}s_k(J(x_k)) \in \mathcal{X}$, and $J(x_k), J(x_k + \sqrt{h}s_k(J(x_k))) \in \mathcal{J}$, thus, there exist $\bar{M}_{J,m-1}, \bar{M}_{s,m-1}$, $\bar{M}^+_{J,m-1}, \bar{M}^+_{s,m-1} \in \mathbb{R}_{\geq 0}$ such that $\|\bar{R}_{J,m-1}(x_0, J(x_0); h)\|_2 \leq \bar{M}_{J,m-1}h$ and so on. Note that $\bar{M}_{J,m-1}$ is derived by (A.27) as

$$\bar{M}_{J,m-1} = L_J^2 \Big(\alpha_2 \sum_{i=0}^{m-2} K_{s,i}L_{s,i} + (\alpha_1 + \alpha_2)^2 \sum_{i=0}^{i-1} K_{s,i}L_{s,i}\Big) + \sqrt{h}M_{m-2} + M_{J,m-1} \tag{A.34}$$

with $L_J, K_{s,i}, L_{s,i}, M_{m-2} \in \mathbb{R}_{\geq 0}$, where $\|\nabla J(x_0)\|_2 \leq L_J$, $\|\nabla s_i(J(x_0))\|_2 \leq K_{s,i}$, $\|\partial s_i/\partial J(x_0)\|_2 \leq L_{s,i}$, and $\|R_{m-2}(x_0, J(x_0); h^{3/2})\|_2 \leq \bar{M}_{m-2}h^{3/2}$. Additionally, by Lemma 3 there exists a $M_{J,m-1} \in \mathbb{R}_{>0}$ such that $\|R_{J,m-1}(x_0, J(x_0); h)\|_2 \leq M_{J,m-1}h$. The constants $\bar{M}_{s,m-1}, \bar{M}^+_{J,m-1}$, $\bar{M}^+_{s,m-1}$ are derived in the same manner via (A.29), (A.31), and (A.33), respectively. Finally, plugging (A.28) and (A.32) in (A.25) yields (A.17) with $k = 0$ and

$$R_{m-1}(x_0, J(x_0); h^{3/2}) = R_{m-2}(x_0, J(x_0); h^{3/2}) + \sqrt{h}\alpha_1\bar{R}_{s,m-1}(x_0, J(x_0); h)$$

$$+ \sqrt{h}\alpha_2\bar{R}^+_{s,m-1}(x_0, J(x_0); h), \tag{A.35}$$

where,

$$\|R_{m-1}(x_0, J(x_0); h^{3/2})\|_2 \leq M_{m-1} h^{3/2} := (M_{m-2} + \alpha_1 \bar{M}_{s,m-1} + \alpha_1 \bar{M}_{s,m-1}^+) h^{3/2}. \quad \text{(A.36)}$$

$\square$

A.3.1 Validation of Conjecture 1

Due to the dependency of $C(m)$ on m in (3.22), the interlacing eigenvalue lemmas, see Lemma 4 and Lemma 5, are not applicable, since the entries of $C(m)$ change with dimension due to $\epsilon(m)$. We verified numerically that the interlacing property (3.25) hold for all $m \leq 10000$. Moreover, in Figure A.1 the interlacing property for $4 \leq m \leq 10$ is visualized.

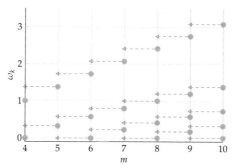

Figure A.1. Illustration of the absolute value of the complex conjugated eigenvalues $\pm \omega_k i$ (•) of $\bar{P}$ in (3.26) w.r.t. m for $k = 1, \ldots, \lceil m/2 \rceil$. The interlacing property (3.25) is apparent and visualized by the arrows; mapping the eigenvalues to the next lower dimension.

Note that m acts as the exploration sequences lengths as introduced in (1.3) and (3.2), where in worst case $m = 4n$ (cf. (3.11)). This implies the property holds for sure for $n \leq 2500$ dimensional problems, which is a very high-dimensional problem for derivative-free optimization algorithms.

A.3.2 Characteristic Polynomial of (3.22)

We propose the characteristic polynomial of the skew-symmetric matrix $C(m)$ in (3.22) as follows:

$$p(\lambda, m) = \lambda^m + \sum_{k=0}^{m/2-1} \left[\left(m(m - 2k)\epsilon(m)^2 + 2(m - 2k)\epsilon(m) \right) \binom{m+1}{m+1-2k} \right.$$
$$\left. + \binom{m}{m - 2k} \right] \lambda^{2k}$$

if m even, and

$$p(\lambda, m) = \lambda^m + \sum_{k=0}^{(m-1)/2-1} \left[\left(m(m-1-2k)\epsilon(m)^2 + 2(m-1-2k)\epsilon(m) \right) \binom{m+1}{m-2k} \right.$$
$$\left. + \binom{m}{m-1-2k} \right] \lambda^{2k+1}$$

if m odd. As pointed out in Section 3.6, this might be a fruitful starting point to prove Conjecture 1.

B
Proofs

B.1 Proof of Theorem 1

The proof utilizes the result of Lemma 6. Consider the $k + m$-th step of the evolution of (3.1) represented by (A.17) with transition map (3.2). Let $w_\ell = [u_\ell^\top \; v_\ell^\top]^\top$ for $\ell = k, \ldots, k + m - 1$ and

$$Y(f(z), g(z)) = [f(z)I \quad g(z)I], \tag{B.1}$$

$$\tilde{Y}(f(z), g(z)) = \left[\tfrac{\partial f}{\partial z}(z)I \quad \tfrac{\partial g}{\partial z}(z)I\right]. \tag{B.2}$$

Then plugging $\{w_\ell\}_{\ell=k}^{k+m-1}$, $Y(f(J(x_k)), g(J(x_k)))$, and $\tilde{Y}(f(J(x_k)), g(J(x_k)))$ in (A.17) yields

$$x_{k+m} = x_k + \sqrt{h}(\alpha_1 + \alpha_2)Y(f(J(x_k)), g(J(x_k))) \sum_{i=k}^{k+m-1} w_i + \mathcal{O}(h^{3/2})$$

$$+ h\tilde{Y}(f(J(x_k)), g(J(x_k))) \left\{ \sum_{i=k}^{k+m-1} \left(\alpha_2 w_i w_i^\top + (\alpha_1 + \alpha_2)^2 \sum_{j=k}^{i-1} w_i w_j^\top \right) \right\}$$

$$\times Y(f(J(x_k)), g(J(x_k)))^\top \nabla J(x_k). \tag{B.3}$$

The term in the curly brackets in (B.3) is $T(W)$ in (3.4); therefore, (3.3) is recovered. $\square$

B.2 Proof of Theorem 2

Let $\delta_0, \ldots, \delta_3 \in \mathbb{R}_{\geq 0}$ with $0 \leq \delta_3 \leq \delta_2 \leq \delta_1 \leq \delta_0$ and $\mathcal{U}_{\delta_0}(x^*), \mathcal{U}_{\delta_1}(x^*), \mathcal{U}_{\delta_2}(x^*), \mathcal{U}_{\delta_3}(x^*) \subseteq \mathbb{R}^n$ are compact convex sets with center point x^* (illustrated in Figure B.1). The proof is separated in:

Step 1) Define a radially unbounded *Lyapunov*-like function $V \in C^1(\mathcal{U}_{\delta_0}(x^*), \mathbb{R}^n)$, such that $V(x_{k+m}) - V(x_k) < 0$ for $\mathcal{U}_{\delta_0}(x^*) \backslash \{\mathcal{U}_{\delta_3}(x^*)\}$ (periodic stability and local attractivity of $\mathcal{U}_{\delta_3}(x^*)$, cf. Moreau and Aeyels (2000); Khalil (2002); Vidyasagar (2002))

Step 2) Practical invariance of $\mathcal{U}_{\delta_1}(x^*)$

Step 3) Convergence of x_k to $\mathcal{U}_{\delta_2}(x^*)$ with $x_0 \in \mathcal{U}_{\delta_1}(x^*)$

Step 1) Consider the algorithm (3.1) with transition map (3.2) and its evolution as given in (3.3) where (3.8) and (3.7) are satisfied. Moreover, δ_0 is chosen such that for every $k \geq 0$ it hold $x_k, x_k + \sqrt{h}s_k(J(x_k)) \in \mathcal{U}_{\delta_0}(x^*)$. Thus, by Lemma 6 there exist a $R_{k+m-1} : \mathbb{R}^n \times \mathbb{R} \times \mathbb{R}_{\geq 0} \to \mathbb{R}^n$ and a constant $M_{k+m-1} \in \mathbb{R}_{\geq 0}$ such that

$$x_{k+m} = x_k - h\nabla J(x_k) + h^{3/2}R_{k+m-1}(x_k, J(x_k); 1) \tag{B.4}$$

with $\|R_{k+m-1}(x_k, J(x_k); 1)\|_2 \leq M_{k+m-1}$. Consider the Lyapunov-like candidate function $V(x_k) = J(x_k) - J(x^*)$ (it holds $V(x^*) = 0$; $V(x) > 0$, $x \in \mathbb{R}^n\backslash\{x^*\}$), then by Lemma 3 there exist a $\bar{R} : \mathbb{R}^n \times \mathbb{R} \times \mathbb{R}^n \times \mathbb{R}_{>0} \to \mathbb{R}$ and a $\bar{M}_{k+m-1} \in \mathbb{R}_{\geq 0}$ such that for all $k \geq 0$ and for every $m \geq 0$

$$\begin{aligned}
V(x_{k+m}) - V(x_k) &= J\left(x_k - h\nabla J(x_k) + h^{3/2}R_{k+m-1}(x_k, J(x_k); 1)\right) - J(x_k) \\
&= -h\nabla J(x_k)^\top \nabla J(x_k) + h^{3/2}\nabla J(x_k)^\top R_{k+m-1}(x_k, J(x_k); 1) \\
&\quad + h^2 \bar{R}_{k+m-1}(x_k, J(x), \nabla J(x_k); 1)
\end{aligned} \tag{B.5}$$

with $\|\bar{R}_{k+m-1}(x_k, J(x_k), \nabla J(x_k); 1)\|_2 \leq \bar{M}_{k+m-1}$ hold. Therefore, for every δ_0 there exists a $M_{\delta_0} \in \mathbb{R}_{\geq 0}$ such that for all $x_k \in \mathcal{U}_{\delta_0}(x^*)$ with $k \geq 0$ and every $m \geq 0$

$$V(x_{k+m}) - V(x_k) \leq -h\|\nabla J(x_k)\|_2^2 + h^{3/2}M_{\delta_0} \tag{B.6}$$

is satisfied. Then under Assumption **[A2]**, there exist $h_1 \in \mathbb{R}_{>0}$ and $\epsilon \in \mathbb{R}_{>0}$, such that for every $h \in \,]0, h_1[$ hold

$$V(x_{k+m}) - V(x_k) < -\epsilon \quad \text{on } \mathcal{L} = \mathcal{L}_0\backslash\{\mathcal{L}_3\}, \tag{B.7}$$

with $\mathcal{L} \supseteq \mathcal{U}_{\delta_0}(x^*)\backslash\{\mathcal{U}_{\delta_3}(x^*)\}$ and sub-level sets $\mathcal{L}_0 \supseteq \mathcal{U}_{\delta_0}(x^*)$ and $\mathcal{L}_3 \subseteq \mathcal{U}_{\delta_3}(x^*)$ of $V(x)$.

Step 2) By Lemma 3, Assumption **[A1]**, and (3.1) with (3.2), it hold that for every $x_k \in \mathcal{U}_{\delta_0}(x^*)$ with $k \geq 0$ and every $\ell \in [0, m]$ there exist a $\hat{R}_{k,\ell} : \mathbb{R}^n \times \mathbb{R} \times \mathbb{R}_{>0} \to \mathbb{R}$ and a $\hat{M}_{k,\ell} \in \mathbb{R}$ such that

$$\begin{aligned}
V(x_{k+\ell}) &= V\left(x_k + \sqrt{h}\alpha_1 s_k\left(J(x_k)\right) + \sqrt{h}\alpha_2 s_k\left(J\left(x_k + \sqrt{h}s_k(J(x_k))\right)\right)\right) \\
&= J(x_k) - J(x^*) + \sqrt{h}\hat{R}_{k,\ell}(x_k, J(x_k); 1) \\
&\leq V(x_k) + \sqrt{h}\hat{M}_{k,\ell}.
\end{aligned} \tag{B.8}$$

Thus, in combination with Step 1) there exists an $h_2 \in \mathbb{R}_{>0}$ such that for all $h \in \,]0, h_2[$ and $x_0 \in \mathcal{U}_{\delta_1}(x^*)$ it follows that $x_k \in \mathcal{U}_{\delta_0}(x^*)$ for all $k \geq 0$.

Step 3) Due to (B.7), it holds that for every initial value $x_0 \in \mathcal{U}_{\delta_1}(x^*)$, there exists a maximum number of iterations $N \in \mathbb{N}$, such that for every $h \in \,]0, \min\{h_1, h_2\}[$, it holds $x_N \in \mathcal{U}_{\delta_3}(x^*)$. Furthermore, similar as in Step 2), there exists an $h_3 \in \mathbb{R}_{>0}$, such that for all $h \in \,]0, h_3[$ and $x_0 \in \mathcal{U}_{\delta_3}(x^*)$ it follows that $x_k \in \mathcal{U}_{\delta_2}(x^*)$ for all $k \geq 0$. Thus, for all $h \in \,]0, \min\{h_1, h_2, h_3\}[$ and if $x_0 \in \mathcal{U}_{\delta_1}(x^*)$, then x_k converges to $\mathcal{U}_{\delta_2}(x^*)$ in N iterations. Figure B.1 illustrates the proof. $\qquad\square$

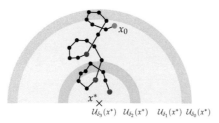

Figure B.1. An illustration of the proof of Theorem 2. Convergence of $x_k \in \mathbb{R}^2$ (•) with initial value x_0 (•) $\in \mathcal{U}_{\delta_1}(x^*) \subseteq \mathcal{U}_{\delta_0}(x^*)$ to $\mathcal{U}_{\delta_2}(x^*)$. For all $k > N$, x_k stays in $\mathcal{U}_{\delta_2}(x^*)$. Note, every m-th step is marked with •.

B.3 Proof of Theorem 3

Since the step sizes h_k is constant for each interval $k \in \{pm, \ldots, p(m+1) - 1\}$ for $p = 0, 1, \ldots$, Lemma 6 holds and a gradient step as in (B.4) is approximated every $k = pm$ with $h = h_{pm}$, thus

$$x_{(p+1)m} = x_{pm} - h_{pm}\nabla J(x_{pm}) + h^{3/2}R_{(p+1)m-1}(x_{pm}, J(x_{pm}); 1). \tag{B.9}$$

Following the same lines as in the proof of Theorem 2 in Appendix B.2 there exists a $M_{(p+1)m-1} \in \mathbb{R}_{\geq 0}$ such that $\|R_{(p+1)m-1}(x_{pm}, J(x_{pm}); 1)\|_2 \leq M_{(p+1)m-1}$. Therefore with Lemma 2.2.1 and Theorem 2.3.1 in Kushner and Clark (2012), the assumption on the step size h_k in Theorem 3, and Appendix B.2, there exists an $h_0^* \in \mathbb{R}_{>0}$ such that for all $h_0 \in]0, h_0^*[$ hold that $x_0 \to x^*$ for $k \to \infty$. □

B.4 Proof of Lemma 1

Condition (3.7) implies

$$w_{m-1} = -\sum_{i=0}^{m-2} w_i. \tag{B.10}$$

Plugging (B.10) in (3.4) yields

$$
\begin{aligned}
T(W) &= \sum_{i=0}^{m-2}\left(\alpha_2 w_i w_i^\top + (\alpha_1 + \alpha_2)^2 \sum_{j=0}^{i-1} w_i w_j^\top\right) + \alpha_2 w_{m-1} w_{m-1}^\top + (\alpha_1 + \alpha_2)^2 \sum_{i=0}^{m-2} w_{m-1} w_i^\top \\
&= \alpha_2 \sum_{i=0}^{m-2}\left(w_i w_i^\top + \sum_{j=0}^{m-2} w_i w_j^\top\right) + (\alpha_1 + \alpha_2)^2 \sum_{i=0}^{m-2}\left(\sum_{j=0}^{i-1} w_i w_j^\top - \sum_{j=0}^{m-2} w_i w_j^\top\right)
\end{aligned}
$$

$$= \alpha_2 \sum_{i=0}^{m-2} \left(w_i w_i^\top + \sum_{j=0}^{m-2} w_i w_j^\top \right) - (\alpha_1 + \alpha_2)^2 \sum_{i=0}^{m-2} \sum_{j=i}^{m-2} w_i w_j^\top$$

$$= (\alpha_2 - (\alpha_1 + \alpha_2)^2) \sum_{i=0}^{m-2} \sum_{j=i}^{m-2} w_i w_j^\top + \alpha_2 \sum_{i=0}^{m-2} \sum_{j=0}^{i} w_i w_j^\top. \tag{B.11}$$

Hence, P in (3.20) is recovered.   □

B.5 Proof of Theorem 4

Consider (3.21) and the singular value decomposition of the exploration sequence matrix

$$W = U \Sigma V^\top \tag{B.12}$$

with

- $U = [a_1 \, b_1 \, \cdots \, a_n \, b_n]$, \hfill (B.13)

- $\Sigma = \begin{bmatrix} \Sigma_0 & 0 \\ 0 & 0 \end{bmatrix}$ with $\Sigma_0 = \mathrm{diag}([\sigma_1 \, \cdots \, \sigma_r])$, \hfill (B.14)

- $V = \begin{bmatrix} \Theta - \epsilon \mathbf{1} \mathbf{1}^\top \Theta & m^{-1/2} \mathbf{1} \\ -\mathbf{1}^\top \Theta + \epsilon (m-1) \mathbf{1}^\top \Theta & m^{-1/2} \end{bmatrix}$ with $\epsilon = (m-1)^{-1}(1 - m^{-1/2})$ and

  $\Theta \in \mathbb{R}^{(m-1) \times (m-1)}$ s.t. $\Theta^\top \Theta = \Theta \Theta^\top = I$. \hfill (B.15)

Hereby, $a_\ell \pm b_\ell i$ with $a_\ell, b_\ell \in \mathbb{R}^{2n}$ for $\ell = 1, \ldots, n$ are the eigenvectors of T_d, $\sigma_1, \ldots, \sigma_r \in \mathbb{R}_{>0}$ are the singular values of W, and $r := \mathrm{rk}(T_d)$. Since U, as defined in (B.13), is constructed by the real and imaginary parts of the eigenvectors of the matrix T_d, U is orthogonal (cf. Horn and Johnson (2012) or Appendix A.2). Moreover,

$$X := U^\top T_d U = \mathrm{diag}([C_1 \, \cdots \, C_n]) \quad \text{with} \quad C_\ell = \begin{bmatrix} \gamma_\ell & -\delta_\ell \\ \delta_\ell & \gamma_\ell \end{bmatrix}, \; \ell = 1, \ldots, n, \tag{B.16}$$

where $\gamma_\ell \pm \delta_\ell i$ are the eigenvalues of T_d with $(2\alpha_2 - (\alpha_1 + \alpha_2)^2)\gamma_\ell \in \mathbb{R}_{\geq 0}$ and $\delta_\ell \in \mathbb{R}_{\geq 0}$. Note that $\gamma_\ell = 0$ and $\delta_\ell = 0$ for $\ell > \lceil r/2 \rceil$.

The orthogonality of V, given in (B.15), is shown by direct evaluations:

$$V^\top V = \begin{bmatrix} \hat{V}_{11} & 0 \\ 0 & 1 \end{bmatrix} \quad \text{with} \tag{B.17}$$

$$\hat{V}_{11} = I + (1 - 2\epsilon m + \epsilon^2 m(m-1)) \Theta^\top \mathbf{1} \mathbf{1}^\top \Theta,$$

$$V V^\top = \begin{bmatrix} \tilde{V}_{11} & \tilde{V}_{12} \\ \tilde{V}_{12}^\top & \tilde{V}_{22} \end{bmatrix} \quad \text{with} \tag{B.18}$$

$$\tilde{V}_{11} = I + (\epsilon^2(m-1) + m^{-1} - 2\epsilon) \mathbf{1} \mathbf{1}^\top$$

$$\tilde{V}_{12} = -(\epsilon^2(m-1)^2 - 2\epsilon(m-1) - m^{-1} + 1) \mathbf{1}$$

$$\tilde{V}_{22} = \epsilon^2(m-1)^3 - 2\epsilon(m-1)^2 + m^{-1} + m - 1$$

where we used the fact that Θ in (B.15) is orthogonal. By plugging $\epsilon = (m-1)^{-1}(1 - m^{-1/2})$ in (B.17) and (B.18), the orthogonality of V, i.e., $VV^\top = V^\top V = I$ is recovered. Now plugging (B.12) in (3.21) associated with (B.16) reveals

$$\begin{bmatrix} \Sigma_0 & 0 \\ 0 & 0 \end{bmatrix} V^\top P V \begin{bmatrix} \Sigma_0 & 0 \\ 0 & 0 \end{bmatrix} = X, \tag{B.19}$$

where

$$Q := V^\top P V = \begin{bmatrix} \tilde{Q} & * \\ * & * \end{bmatrix} \text{ with } \tilde{Q} = \Theta^\top \tilde{P} \Theta \text{ and} \tag{B.20}$$

$$\tilde{P} = \left(P - \epsilon(\mathbb{1}\mathbb{1}^\top P + P\mathbb{1}\mathbb{1}^\top) + \epsilon^2 \mathbb{1}\mathbb{1}^\top P\mathbb{1}\mathbb{1}^\top \right)_{1:m-1} \tag{B.21}$$

with $\tilde{P} \in \mathbb{R}^{(m-1)\times(m-1)}$, which can be written as

$$\tilde{P} = \left(\alpha_2 - \frac{1}{2}(\alpha_1 + \alpha_2)^2 \right)\left(I + (\underbrace{m(m-1)\epsilon^2 - 2m\epsilon^2 + 1}_{=0})\mathbb{1}\mathbb{1}^\top \right)$$

$$+ \frac{1}{2}(\alpha_1 + \alpha_2)^2 \begin{bmatrix} 0 & d_1 & d_2 & \dots & d_{m-2} \\ -d_1 & \ddots & \ddots & \ddots & \vdots \\ -d_2 & \ddots & \ddots & \ddots & d_2 \\ \vdots & \ddots & \ddots & \ddots & d_1 \\ -d_{m-2} & \dots & -d_2 & -d_1 & 0 \end{bmatrix} \tag{B.22}$$

with $d_i = 2i\epsilon - 1$ for $i = 1, \dots, m-2$ and ϵ defined as in (B.15), where $\tilde{P}\tilde{P}^\top - \tilde{P}^\top \tilde{P} = 0$, thus $\tilde{P}$ normal. More precise, $\tilde{P}$ has complex conjugated eigenvalues $\mu_\ell \pm \omega_\ell i$ with $\ell = 1, \dots, \lceil (m-1)/2 \rceil$ where $\mu_\ell = \mu = \alpha_2 - 1/2(\alpha_1 + \alpha_2)^2$ and the skew-symmetric part is a Toeplitz matrix (second line of (B.22)).

Eventually, (B.19) impose the conditions

$$X_{1:r} = \Sigma_0 \tilde{Q}_{1:r} \Sigma_0, \tag{B.23}$$

$$X_{r+1:n} = 0. \tag{B.24}$$

Then, (B.24) holds, since there exist $n - r - 1$ eigenvalues identical to zero and U can be ordered accordingly. Additionally, let Θ in (B.15) be of the form such that

$$\tilde{Q}_{1:r} = \text{diag}([D_1 \cdots D_{\lceil r/2 \rceil}])$$

$$\text{with } D_\ell = \begin{bmatrix} \mu & -\hat{\omega}_\ell \\ \hat{\omega}_\ell & \mu \end{bmatrix}, \ell = 1, \dots, \lceil r/2 \rceil \tag{B.25}$$

holds, where the imaginary part of the eigenvalues of the principal submatrix $\tilde{Q}_{1:r}$ of $\tilde{Q}$ in (B.20) is denoted by $\pm \hat{\omega}_k i$ for $k = 1, \dots, \lceil r/2 \rceil$.

Then (B.23) implies for $k = 1, \ldots, \lceil r/2 \rceil$

$$\mu\sigma_{2k-1}^2 = \mu\sigma_{2k}^2 = \gamma_k \quad \text{and} \tag{B.26}$$

$$\sigma_{2k-1}\sigma_{2k}\hat{\omega}_k = \delta_k. \tag{B.27}$$

In the case of $2\alpha_2 - (\alpha_1 + \alpha_2)^2 = 0$, T_d and $\tilde{P}$ are skew-symmetric due to the assumptions in Theorem 4 and (B.22), respectively. Hence, $\gamma_\ell = 0$ for $\ell = 1, \ldots, n$, and $\mu = 0$, which implies that (B.26) is satisfied. Equation (B.27) is satisfied for $k = 1, \ldots, \lceil r/2 \rceil$ with

$$\sigma_{2k} = \delta_k \omega_k^{-1} \sigma_{2k-1}^{-1} \quad \text{and} \quad \sigma_{2k-1} \in \mathbb{R}_{>0} \tag{B.28}$$

where $m = r + 1$ and therefore $\hat{\omega}_k = \omega_k$. Then Θ is constructed as orthogonal transformation matrix of $\tilde{P}$ (similar to U w.r.t. T_d).

In the case $2\alpha_2 - (\alpha_1 + \alpha_2)^2 \neq 0$, (B.26) and (B.27) together yield

$$\sigma_{2k-1}^2 = \sigma_{2k}^2 = \frac{\delta_k}{\hat{\omega}_k} = \frac{\gamma_k}{\mu}, \tag{B.29}$$

for $k = 1, \ldots, \lceil r/2 \rceil$, hence

$$\hat{\omega}_k = \frac{\delta_k}{\gamma_k}\mu \tag{B.30}$$

has to be satisfied. Note that δ_k, γ_k, and μ are specified by T_d, α_1, and α_2, and $\mu/\gamma_k \geq 0$ due to the positive definiteness condition in Theorem 4. Applying Lemma 5 to $\tilde{P}$ in (B.21) implies that there exists a Θ such that (B.25) and the interlacing property

$$\omega_k \geq \hat{\omega}_k \geq \omega_{\lceil (m-1)/2 \rceil - \lceil r/2 \rceil + k}, \; k = 1, \ldots, \lceil r/2 \rceil \tag{B.31}$$

holds, where $\hat{\omega}_k$ can be chosen in the given intervals. Note that Lemma 5 can be applied to the normal matrix $\tilde{P}$, due to the decomposition of a scaled unit and skew-symmetric matrix. W.l.o.g., δ_k/γ_k and $\hat{\omega}_k$ are sorted in decreasing order. Then by applying Conjecture 1 (cf. Remark 2) to $\tilde{P}$ in (B.21) successively, there exists a $m \geq r + 1$ such that (B.31) with (B.30) for all $k = 1, \ldots, \lceil r/2 \rceil$ holds. $\qquad\Box$

B.6 Proof of Lemma 2

The entry in the p-th row and q-th column of the matrix $T(W)$ in (3.3) with $[\alpha_1 \, \alpha_2] = [1/2 \, 1/2]$ is given by

$$T(W)_{pq} = \sum_{i=0}^{m-1} \frac{1}{2} e_p^\top w_i w_i^\top e_q + \sum_{j=0}^{i-1} e_p^\top w_i w_j^\top e_q. \tag{B.32}$$

Condition $W\mathbf{1} = 0$ implies $w_{m-1} = -\sum_{i=0}^{m-1} w_i$ such that we have

$$
\begin{aligned}
T(W) &= \sum_{i=0}^{m-2} \left(\frac{1}{2} w_i w_i^\top + \sum_{j=0}^{i-1} w_i w_j^\top \right) + \frac{1}{2} w_{m-1} w_{m-1}^\top + \sum_{j=0}^{m-2} w_{m-1} w_j^\top \\
&= \sum_{i=0}^{m-2} \left(\frac{1}{2} w_i w_i^\top + \sum_{j=0}^{i-1} w_i w_j^\top + \frac{1}{2} \sum_{j=0}^{m-2} w_i w_j^\top - \sum_{j=0}^{m-2} w_i w_j^\top \right) \\
&= \frac{1}{2} \sum_{i=0}^{m-2} \left(w_i w_i^\top + \sum_{j=0}^{i-1} w_i w_j^\top - \sum_{j=i}^{m-2} w_i w_j^\top \right) \\
&= \frac{1}{2} \sum_{i=0}^{m-2} \left(\sum_{j=0}^{i-1} w_i w_j^\top - \frac{1}{2} \sum_{j=i+1}^{m-2} w_i w_j^\top \right) \\
&= \frac{1}{2} \sum_{i=1}^{m-2} \sum_{j=0}^{i-1} (w_i w_j^\top - w_j w_i^\top),
\end{aligned}
\tag{B.33}
$$

where the last line of (B.33) is obtained by the reindexing

$$
\begin{aligned}
\sum_{i=0}^{m-2} \sum_{j=i+1}^{m-2} w_i w_j^\top &= w_0 (w_1 + w_2 + \cdots + w_{m-2})^\top \\
&\quad + w_1 (w_2 + w_3 + \cdots + w_{m-2})^\top \\
&\quad + \ldots \\
&\quad + w_{m-3} w_{m-2}^\top = \sum_{i=1}^{m-2} \sum_{j=0}^{i-1} w_j w_i^\top.
\end{aligned}
\tag{B.34}
$$

Hence, it follows for the entry of $T(W)$ in the p-th row and q-th column

$$
T(W)_{pq} = \frac{1}{2} \sum_{i=1}^{m-2} \sum_{j=0}^{i-1} e_p^\top w_i w_j^\top e_q - e_p^\top w_j w_i^\top e_q.
\tag{B.35}
$$

On the other hand, the net area formula of a n-sided polygon with corner points $\{x_{p,i}, y_{q,i}\}_{i=1}^n$ is given as

$$
A_{pq} = \frac{1}{2} \sum_{i=0}^{n-1} x_{p,i+1} y_{q,i} - x_{p,i} y_{q,i+1}.
\tag{B.36}
$$

Plugging (3.29) with $n = m$ in (B.36) and using $\sum_{i=0}^{m-1} w_i = 0$ yields

$$
\begin{aligned}
A_{pq} &= \frac{1}{2} \sum_{i=0}^{m-1} \left(\sum_{j=0}^{i} e_p^\top w_j \right) \left(\sum_{j=0}^{i-1} w_j^\top e_q \right) - \left(\sum_{j=0}^{i-1} e_p^\top w_j \right) \left(\sum_{j=0}^{i} w_j^\top e_q \right) \\
&= \frac{1}{2} \sum_{i=1}^{m-2} e_p^\top w_i \left(\sum_{j=0}^{i-1} w_j^\top e_q \right) - \left(\sum_{j=0}^{i-1} e_p^\top w_j \right) w_i^\top e_q.
\end{aligned}
\tag{B.37}
$$

Consequently, $A_{pq} = T(W)_{pq}$ holds. $\qquad\square$

B.7 Proof of Theorem 5

Let $2\alpha_2 - (\alpha_1 + \alpha_2)^2 = 0$ and $T(W) = T_d$ is skew-symmetric and partitioned as in (3.9). Then condition (3.8) with $T_{12} = -T_{21}^\top = -R \in \mathbb{R}^{n \times n}$, where R is arbitrary, reads

$$-I = g'(z)f(z)R^\top - f'(z)g(z)R + f'(z)f(z)T_{11} + g'(z)g(z)T_{22}, \ \forall z \in \mathbb{R}. \tag{B.38}$$

In the sequel, all terms depending on z must hold for all $z \in \mathbb{R}$. Since, $T_{11} = -T_{11}^\top$, $T_{22} = -T_{22}^\top$, w.l.o.g. we can express R as $R = I + \tilde{R}$, where $\text{diag}(\tilde{R}) = 0$. Thus, it must hold

$$g'(z)f(z) - f'(z)g(z) = -1, \tag{B.39}$$

which is satisfied by the gradient-generating functions (Grushkovskaya et al., 2018, Theorem 1)

$$g(z) = -f(z) \int \frac{1}{f^2(z)} dz. \tag{B.40}$$

Assume that gradient-generating functions f and g as in (B.40) satisfy (B.39), then condition (B.38) translates into

$$f'(z)f(z)T_{11} + g'(z)g(z)T_{22} = (f'(z)g(z) + g'(z)f(z))\tilde{R}, \ \forall z \in \mathbb{R}, \tag{B.41}$$

implying that $\tilde{R} = -\tilde{R}^\top$ holds. Next we consider three cases:

Case 1: $\tilde{R} = 0$. Thus,

$$f'(z)f(z)T_{11} + g'(z)g(z)T_{22} = 0. \tag{B.42}$$

Clearly, (B.42) is satisfied by $T_{11} = T_{22} = 0$ with f arbitrary while satisfying g in (B.40), i.e., (3.33) results.

For the sub-case $a^{-1}T_{11} = b^{-1}T_{22} =: Q$ arbitrary skew-symmetric with $a, b \in \mathbb{R}_{>0}$, f and g have to satisfy $af'(z)f(z) + bg'(z)g(z) = 0$, i.e., w.l.o.g. $af^2(z) + bg^2(z) = 1$. Accordingly, with (B.40) and $y'(z) = f^{-2}(z)$ it yields

$$y'(z) = a + by^2(z). \tag{B.43}$$

The unique solution of (B.43) with $\phi \in \mathbb{R}$ is

$$y(z) = \sqrt{\frac{a}{b}} \tan\left(\sqrt{ab}z + \phi\right); \tag{B.44}$$

therefore, with the definition of $y(z)$ and (B.40) one reveals (3.34).

Repeating the above calculations for $a^{-1}T_{11} = -b^{-1}T_{22} =: Q$ arbitrary skew-symmetric yields

$$y(z) = \sqrt{\frac{a}{b}} \tanh\left(\sqrt{ab}z + \phi\right); \tag{B.45}$$

therefore, with the definition of $y(z)$ and (B.40) one reveals (3.35).

The remaining sub-case $T_{11} \neq \pm a T_{22}$ implies that

$$f'(z)f(z)T_{11} = 0 \text{ and } g'(z)g(z)T_{22} = 0, \tag{B.46}$$

since $f'(z)f(z)T_{11} + g'(z)g(z)T_{22} = 0$ must hold. If $f'(z)f(z) = 0$ and $T_{11} = Q$ arbitrary skew-symmetric, i.e., $f^2(z) = a$ with $a \in \mathbb{R}_{>0}$, it implies that $f(z) = \sqrt{a}$. Thus, with (B.40), $g'(z)g(z) \neq 0$ for all $z \in \mathbb{R}$ yields $T_{22} = 0$, i.e., (3.36) results.

The same argumentation holds for $g'(z)g(z) = 0$ with arbitrary skew-symmetric T_{22} such that (3.37) is recovered.

The circumstance that $T_{11} \neq T_{22}$ with $T_{11} \neq 0$ and $T_{22} \neq 0$ is not valid due to (B.40) and (B.46). Specifically, $f'(z)f(z) = 0$ and $g'(z)g(z) = 0$ must hold; obviously, based on the above cases, $f(z) = \sqrt{a}$ and $g(z) = \sqrt{a}$ are in conflict with (B.39).

Case 2: $f'(z)g(z) + g'(z)f(z) = 0$ for all $z \in \mathbb{R}$. Thus,

$$f'(z)f(z)T_{11} + g'(z)g(z)T_{22} = 0 \text{ and} \tag{B.47}$$

$$f'(z)g(z) + g'(z)f(z) = 0 \tag{B.48}$$

must be satisfied. Clearly, (B.47) is satisfied by $T_{11} = T_{22} = 0$, where (B.48) implies $-af(z)g(z) = 1$ with $a \in \mathbb{R}_{>0}$. Accordingly, with (B.40) and $y'(z) = f^{-2}(z)$ it yields

$$y'(z) = ay(z). \tag{B.49}$$

The unique solution of (B.49) with $c \in \mathbb{R}$ is

$$y(z) = e^{az} + c; \tag{B.50}$$

therefore, with the definition of $y(z)$ and (B.40) one gets (3.38).

The sub-cases where $a^{-1}T_{11} = b^{-1}T_{22} =: Q$ arbitrary skew-symmetric or $T_{11} \neq \pm a T_{22}$ as discussed for Case 1 are not valid. With the same approach as above, i.e., $y'(z) = f^{-2}$ it yields $y'(z) = 0$; therefore, no solution for $f(z)$ (and $g(z)$) can be found or (B.47) and (B.48) are not satisfied as discussed in the last paragraph of Case 1, respectively.

Case 3: $\tilde{R} = -\tilde{R}^\top \neq 0$. Thus,

$$f'(z)f(z)T_{11} + g'(z)g(z)T_{22} = (f'(z)g(z) + g'(z)f(z))\tilde{R}. \tag{B.51}$$

Clearly, $T_{11} = T_{22} = 0$ is not valid, since $(f'(z)g(z) + g'(z)f(z))\tilde{R} \neq 0$ in this last case.

For the sub-case $a^{-1}T_{11} = b^{-1}T_{22} =: Q = -Q^\top \neq 0$ with $a, b \in \mathbb{R}_{>0}$, it must hold that $a/2 f'(z)f(z) + b/2 g'(z)g(z) = c(f'(z)g(z) + g'(z)f(z))$ and $\tilde{R} = cQ$ with $c \in \mathbb{R} \backslash \{0\}$, i.e., w.l.o.g. $a/2 f^2(z) + b/2 g^2(z) - cf(z)g(z) = 1$. Accordingly, with (B.40) and $y'(z) = f^{-2}(z)$ it yields

$$y'(z) = \frac{a}{2} + cy(z) + \frac{b}{2}y^2(z). \tag{B.52}$$

The unique solution of (B.52) with $\phi \in \mathbb{R}$ is

$$y(z) = \frac{\sqrt{ab - c^2}}{b} \tan \left(\sqrt{ab - c^2}z + \phi \right);$$ (B.53)

therefore, with the definition of $y(z)$ and (B.40) one reveals (3.39).

The remaining sub-case $T_{11} \neq aT_{22}$, implies that

$$f'(z)f(z)T_{11} - (f'(z)g(z) + g'(z)f(z))\bar{R} = 0 \quad \text{and} \quad g'(z)g(z)T_{22} = 0, \quad \text{or} \quad \text{(B.54)}$$
$$g'(z)g(z)T_{22} - (f'(z)g(z) + g'(z)f(z))\bar{R} = 0 \quad \text{and} \quad f'(z)f(z)T_{11} = 0, \quad \text{(B.55)}$$

since (B.51) must hold. However, we show in the sequel that (B.54) and (B.55) lead to no new solution or is not valid, respectively.

For (B.54), $a^{-1}T_{11} = c^{-1}\bar{R} =: Q = -Q^\top \neq 0$ with $a \in \mathbb{R}_{>0}$ and $c \in \mathbb{R} \backslash \{0\}$, it must hold that $af'(z)f(z) = c(f'(z)g(z) + g'(z)f(z))$, i.e., w.l.o.g.

$$\frac{a}{2}f^2(z) - cf(z)g(z) = 1.$$ (B.56)

Accordingly, with (B.40) and $y'(z) = f^{-2}(z)$ it yields

$$y'(z) = \frac{a}{2} + cy(z).$$ (B.57)

The unique solution of (B.55) with $d \in \mathbb{R}$ is

$$y(z) = de^{cz} - \frac{a}{2c};$$ (B.58)

therefore, with the definition of $y(z)$ and (B.40) one gets $f(z) = (cd)^{-1/2}\exp(-c/2 z)$ and $g(z) = -(d/c)^{-1/2}\exp(c/2 z)$ such that $T_{22} = 0$. However, f, g satisfying (B.56) for all $z \in \mathbb{R}$ only for $a = 0$, hence, $T_{11} = 0$, i.e, the same result as (3.38).

For (B.55), $b^{-1}T_{11} = c^{-1}\bar{R} =: Q = -Q^\top \neq 0$ with $a \in \mathbb{R}_{>0}$ and $c \in \mathbb{R} \backslash \{0\}$, it must hold that $bg'(z)g(z) = c(f'(z)g(z) + g'(z)f(z))$, i.e., w.l.o.g.

$$\frac{b}{2}g^2(z) - cf(z)g(z) = 1.$$ (B.59)

Following the procedure above yield $f(z) = (2)^{-1/2}c^{-1}(b\exp(c/2 z + d/2) - \exp(-c/2 z - d/2))$ and $g(z) = 2^{1/2}b^{-1}\exp(-c/2z - d/2)$ with $d \in \mathbb{R}$. However, (B.59) is only satisfied for $d = -cz - \ln(b)$ which leads to $f(z) = 0$; therefore, no valid solution.

Notice that every feasible structure of the skew-symmetric matrix T_d is discussed above case by case, and the differential equations arising in the analysis are solved uniquely. Thus, we believe that the list of triples (T_d, f, g) in Theorem 5 for $2\alpha_2 - (\alpha_1 + \alpha_2)^2 = 0$ and T_d skew-symmetric is essentially exhaustive, save for some scaled version of the presented cases.   □

B.8 Proof of Theorem 6

Let T_d be normal and $(2\alpha_2 - (\alpha_1 + \alpha_2)^2)(T_d + T_d^\top)$ be positive definite and let $T(W) = T_d$ be partitioned as in (3.9). Then equation (3.8) reads

$$-I = f'(z)f(z)T_{11} + f'(z)g(z)T_{12} + g'(z)f(z)T_{21} + g'(z)g(z)T_{22}, \forall z \in \mathbb{R}. \tag{B.60}$$

In the sequel, all terms depending on z must hold for all $z \in \mathbb{R}$. Choosing T_d as in (3.40), it holds that $(2\alpha_2 - (\alpha_1 + \alpha_2)^2)(T_d + T_d^\top)$ is positive definite, which yields

$$-1 = a\Big(f'(z)f(z) + g(z)'g(z)\Big) + g'(z)f(z) - f'(z)g(z). \tag{B.61}$$

This equation has been considered in Feiling et al. (2019). We refer to the proof of (Feiling et al., 2019, Theorem 1) for the derivation of f and g as specified in (3.40).

Case (3.41) is analogous to (3.34). First, note that $(2\alpha_2 - (\alpha_1 + \alpha_2))(T_d + T_d^\top)$ is positive definite and normal with the given T_d in (3.41), and $(2\alpha_2 - (\alpha_1 + \alpha_2))(Q + Q^\top)$ is positive definite and normal, since

$$T_d T_d^\top - T_d^\top T_d = \begin{bmatrix} QQ^\top - Q^\top Q & 0 \\ 0 & QQ^\top - Q^\top Q \end{bmatrix} \tag{B.62}$$

and

$$T_d + T_d^\top = \begin{bmatrix} Q + Q^\top & 0 \\ 0 & Q + Q^\top \end{bmatrix}, \tag{B.63}$$

such that the real part of the eigenvalues of Q is identical to that of T_d; therefore, the definiteness property is conserved. Hence, the derivation of the gradient-generating functions f and g in this case goes along the lines of argument as used in the proof of Theorem 5 in Appendix B.7, specifically for the case (3.34), i.e., (B.43) and (B.44). □

The structure of normal matrices T_d come with more degrees of freedom compared to T_d skew-symmetric as in Theorem 5; thus, the two cases listed in Theorem 6 are not exhaustive.

C

Additional Material on Numerical Results

C.1 Parameter Study

In this section we provide visualizations of the data points as described in Section 4.3. First, data points k^* (steps to converge into $\mathcal{U}_{\sqrt{h}}(x^*)$) in Figure C.1 – Figure C.7 and Figure C.13 – Figure C.17 for the algorithm characteristics [C2] and [C1] (as defined on p. 40), respectively. Second, data points ρ (averaged gradient approximation error) in Figure C.2 – Figure C.8 and Figure C.14 – Figure C.18 for the algorithm characteristics [C2] and [C1], respectively. We discuss the visualizations in Section 4.3 and do not further comment them in this section.

C.1.1 Algorithm Setups for Section 4.3

In the following we list the algorithm setups for which the quantitative parameter study (as described in Section 4.3) is carried out. Each setup is tested with the singular values $\sigma = \{\sigma_1, \ldots, \sigma_{\mathrm{rk}(T_d)}\}$ of W chosen as σ^*, $0.5\sigma^*$ $0.9\sigma^*$, $1.1\sigma^*$, $\sigma^{(1)} = \{1, \ldots, 1\}$, and $\sigma^{(2)} = \{2, \ldots, 2\}$, where σ^* are the singular values corresponding to the minimal sequence length m w.r.t. T_d and α_1, α_2. The step size is set to $h = 0.05$ and the initial values x_0 as described in Section 4.3.

Algorithm Setup 10. McOpt with char. [C2] corresponding to (3.33) and (3.34) with

- Generating functions: $f(z) = a^{-1/2} \sin(\sqrt{ab}z)$, $g(z) = b^{-1/2} \cos(\sqrt{ab}z)$

- $T_d = \begin{bmatrix} adQ & -I \\ I & bdQ \end{bmatrix} \in \mathbb{R}^{4 \times 4}$ and $Q = Q_1 = 0$ or $Q = Q_2 = \begin{bmatrix} 0 & -1 \\ 1 & 0 \end{bmatrix}$

where $a, b \in \{0.1, 0.5, 1, 2, 10\}$, and $d \in \{0.2, 1, 3\}$. Step size h, initial values x_0, and singular values of W are chosen as described in the beginning of this subsection. •

Algorithm Setup 11. McOpt with char. [C2] corresponding to (3.33) and (3.35) with

- Generating functions: $f(z) = a^{-1/2} \sinh(\sqrt{ab}z)$, $g(z) = -b^{-1/2} \cosh(\sqrt{ab}z)$

- $T_d = \begin{bmatrix} adQ & -I \\ I & -bdQ \end{bmatrix} \in \mathbb{R}^{4\times4}$ and $Q = Q_1 = 0$ or $Q = Q_2 = \begin{bmatrix} 0 & -1 \\ 1 & 0 \end{bmatrix}$

where $a, b \in \{0.01, 0.05, 0.1, 0.2, 1\}$, and $d \in \{0.2, 1, 3\}$. Step size h, initial values x_0, and singular values of W are chosen as described in the beginning of this subsection.   •

Algorithm Setup 12. McOpt with char. **[C2]** corresponding to (3.33) and (3.36) with

- Generating functions: $f(z) = \sqrt{a}$, $g(z) = a^{-1/2}z$

- $T_d = \begin{bmatrix} dQ & -I \\ I & 0 \end{bmatrix} \in \mathbb{R}^{4\times4}$ and $Q = Q_1 = 0$ or $Q = Q_2 = \begin{bmatrix} 0 & -1 \\ 1 & 0 \end{bmatrix}$

where $a \in \{0.1, 0.5, 1, 2, 10\}$, and $d \in \{0.2, 1, 3\}$. Step size h, initial values x_0, and singular values of W are chosen as described in the beginning of this subsection.   •

Algorithm Setup 13. McOpt with char. **[C2]** corresponding to (3.33) and (3.38) with

- Generating functions: $f(z) = a^{-1/2}e^{-az/2}$, $g(z) = -a^{-1/2}e^{az/2}$

- $T_d = \begin{bmatrix} 0 & -I - dQ \\ I - dQ & 0 \end{bmatrix} \in \mathbb{R}^{4\times4}$ and $Q = Q_1 = 0$ or $Q = Q_2 = \begin{bmatrix} 0 & -1 \\ 1 & 0 \end{bmatrix}$

where $a \in \{0.01, 0.05, 0.1, 0.2, 1\}$, and $d \in \{0.2, 1, 3\}$. Step size h, initial values x_0, and singular values of W are chosen as described in the beginning of this subsection.   •

Algorithm Setup 14. McOpt with char. **[C2]** corresponding to (3.33) and (3.39) with

- Generating functions: $f(z) = \sqrt{b}(ab - c^2)^{-1/2}\sin(\sqrt{ab - c^2}z)$,
  $g(z) = -b^{-1/2}\cos(\sqrt{ab - c^2}z)$

- $T_d = \begin{bmatrix} ad/2Q & -I - cdQ \\ I - cdQ & bd/2Q \end{bmatrix} \in \mathbb{R}^{4\times4}$ and $Q = Q_1 = 0$ or $Q = Q_2 = \begin{bmatrix} 0 & -1 \\ 1 & 0 \end{bmatrix}$

where $a, b \in \{0.1, 0.5, 1, 2, 10\}$, $c \in \{0.2, 0.5, 1\}$, and $d \in \{0.2, 1, 3\}$. Step size h, initial values x_0, and singular values of W are chosen as described in the beginning of this subsection.   •

Algorithm Setup 15. McOpt with char. **[C2]** corresponding to (3.33) with

- Generating functions: $f(z) = a^{-1/2}\sqrt{z}\sin(\mu\sqrt{ab}\ln(z))$,
  $g(z) = b^{-1/2}\sqrt{z}\cos(\mu\sqrt{ab}\ln(z))$

- $T_d = \begin{bmatrix} 0 & -I \\ I & 0 \end{bmatrix} \in \mathbb{R}^{4\times4}$

where $a, b \in \{0.1, 0.5, 1, 2, 10\}$ and $\mu \in \{0.5, 1, 2, 5\}$. Step size h, initial values x_0, and singular values of W are chosen as described in the beginning of this subsection.   •

Algorithm Setup 16. McOpt with char. **[C2]** corresponding to (3.33) where

- Generating functions: $f(z) = a^{-1/2}z\sin(\mu\sqrt{ab}z^{-1})$, $g(z) = b^{-1/2}z\cos(\mu\sqrt{ab}z^{-1})$

- $T_d = \begin{bmatrix} 0 & -I \\ I & 0 \end{bmatrix} \in \mathbb{R}^{4\times4}$

where $a, b \in \{0.1, 0.5, 1, 2, 10\}$ and $\mu \in \{0.5, 1, 2, 5\}$. Step size h, initial values x_0, and singular values of W are chosen as described in the beginning of this subsection. •

Algorithm Setup 17. McOpt with char. **[C1]** corresponding to (3.41) with

- Generating functions: $f(z) = b^{-1/2}\sin(bz)$, $g(z) = b^{-1/2}\cos(bz)$

- $T_d = \begin{bmatrix} Q & -I \\ I & Q \end{bmatrix} \in \mathbb{R}^{4\times4}$ with $Q = Q_1$ as described in Corollary 2 and

$$Q = Q_2 = d\begin{bmatrix} -0.9 & 0.1 \\ 0.1 & -0.9 \end{bmatrix}$$

where $-b \in \{0.1, 0.3, 0.5, 0.8, 1, 2, 3, 5, 10\}$ and $d \in \{0.2, 1, 3\}$. For the construction of Q_2 the singular values of W are chosen as follows: $\sigma^{(1)} = \{2.481, 2.481, 1.205, 1.205\}$, $\sigma^{(2)} = 0.5\{1, 1, 1, 1\}$, $\sigma^{(3)} = 0.75\{1, 1, 1, 1\}$, $\sigma^{(4)} = \{1, 1, 1, 1\}$, $\sigma^{(5)} = 2\{1, 1, 1, 1\}$, and $\sigma^{(6)} = 3\{1, 1, 1, 1\}$ (Note that $\sigma^{(1)}$ is chosen such that $m = 5$ is minimal for the given T_d and α_1, α_2 setting). Step size h, and initial values x_0 are chosen as described in the beginning of this subsection. •

Algorithm Setup 18. McOpt with char. **[C1]** corresponding to (3.40) with

- Generating functions: $f(z) = b^{-1/2}\sqrt{z}\sin(a/2\ln(b^{-1}z) + b\mu\ln(z))$, $g(z) = b^{-1/2}\sqrt{z}\cos(a/2\ln(b^{-1}z) + b\mu\ln(z))$

- $T_d = \begin{bmatrix} aI & -I \\ I & aI \end{bmatrix} \in \mathbb{R}^{4\times4}$

where $-a, b \in \{0.1, 0.5, 1, 2, 10\}$, and $\mu \in \{0.5, 1, 2, 5, 10\}$. Step size h, initial values x_0, and singular values of W are chosen as described in the beginning of this subsection. •

Algorithm Setup 19. McOpt with char. **[C1]** corresponding to (3.40) with

- Generating functions: $f(z) = b^{-1/2}z\sin(-a/2\ln(b^{-1}z) + b\mu z^{-1})$, $g(z) = -b^{-1/2}z\cos(-a/2\ln(b^{-1}z) + b\mu z^{-1})$

- $T_d = \begin{bmatrix} aI & -I \\ I & aI \end{bmatrix} \in \mathbb{R}^{4\times4}$

where $-a, b \in \{0.1, 0.5, 1, 2, 10\}$, and $\mu \in \{0.5, 1, 2, 5, 10\}$. Step size h, initial values x_0, and singular values of W are chosen as described in the beginning of this subsection. •

C.1.2 Convergence and Gradient Approximation Error Data Points

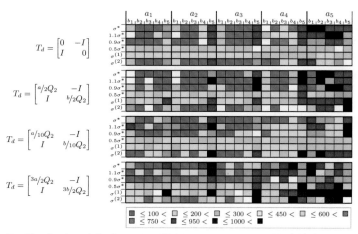

Figure C.1. Visualization of the data points k^* (steps to converge into $\mathcal{U}_{\sqrt{h}}(x^*)$) for Algorithm Setup 10 (see Appendix C.1.1) with T_d as specified applied to (4.4).

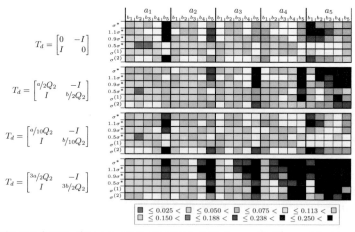

Figure C.2. Visualization of the data points ρ (averaged gradient approximation error) for Algorithm Setup 10 (see Appendix C.1.1) with T_d as specified applied to (4.4).

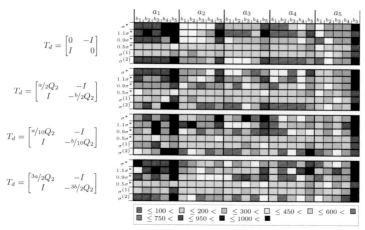

Figure C.3. Visualization of the data points k^* (steps to converge into $\mathcal{U}_{\sqrt{h}}(x^*)$) for Algorithm Setup 11 (see Appendix C.1.1) with T_d as specified applied to (4.4).

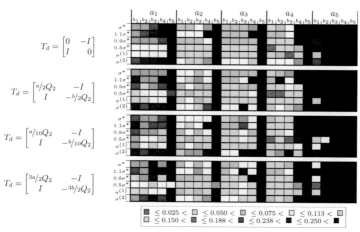

Figure C.4. Visualization of the data points ρ (averaged gradient approximation error) for Algorithm Setup 11 (see Appendix C.1.1) with T_d as specified applied to (4.4).

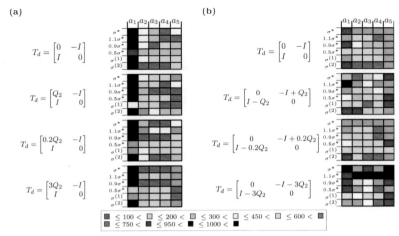

Figure C.5. Visualization of the data points k^* (steps to converge into $\mathcal{U}_{\sqrt{h}}(x^*)$) for **Left (a):** Algorithm Setup 12; **Right (b):** Algorithm Setup 13; with T_d as specified applied to (4.4) (see Appendix C.1.1).

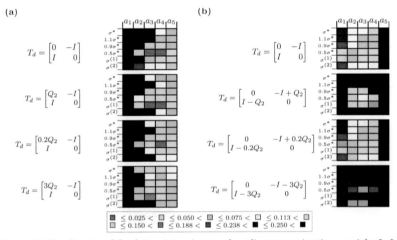

Figure C.6. Visualization of the data points ρ (averaged gradient approximation error) for **Left (a):** Algorithm Setup 12; **Right (b):** Algorithm Setup 13; with T_d as specified applied to (4.4) (see Appendix C.1.1)

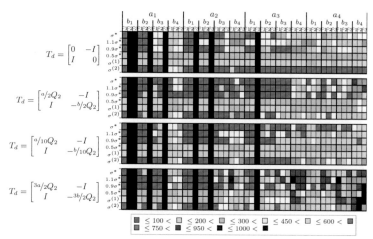

Figure C.7. Visualization of the data points k^* (steps to converge into $\mathcal{U}_{\sqrt{h}}(x^*)$) for Algorithm Setup 14 (see Appendix C.1.1) with T_d as specified applied to (4.4).

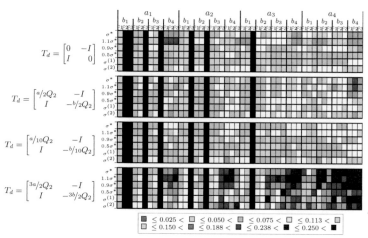

Figure C.8. Visualization of the data points ρ (averaged gradient approximation error) for Algorithm Setup 14 (see Appendix C.1.1) with T_d as specified applied to (4.4).

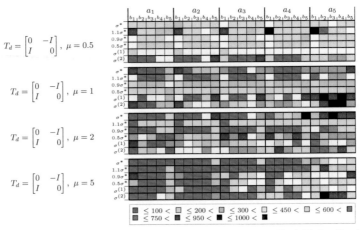

Figure C.9. Visualization of the data points k^* (steps to converge into $\mathcal{U}_{\sqrt{h}}(x^*)$) for Algorithm Setup 15 (see Appendix C.1.1) with T_d as specified applied to (4.4).

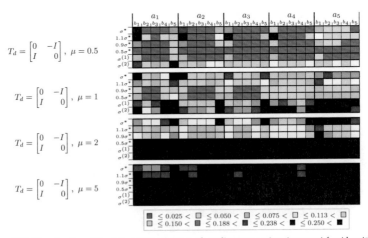

Figure C.10. Visualization of the data points ρ (averaged gradient approximation error) for Algorithm Setup 15 (see Appendix C.1.1) with T_d as specified applied to (4.4).

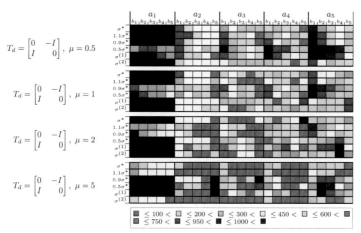

Figure C.11. Visualization of the data points k^* (steps to converge into $\mathcal{U}_{\sqrt{h}}(x^*)$) for Algorithm Setup 16 (see Appendix C.1.1) with T_d as specified applied to (4.4).

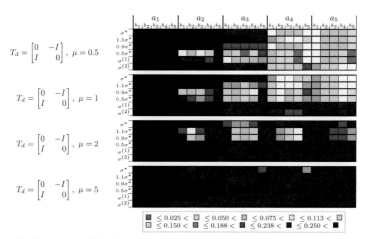

Figure C.12. Visualization of the data points ρ (averaged gradient approximation error) for Algorithm Setup 16 (see Appendix C.1.1) with T_d as specified applied to (4.4).

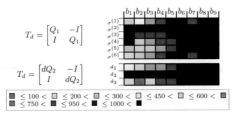

Figure C.13. Visualization of the data points k^* (steps to converge into $\mathcal{U}_{\sqrt{h}}(x^*)$) for Algorithm Setup 17 (see Appendix C.1.1) with T_d as specified applied to (4.4).

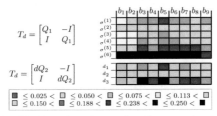

Figure C.14. Visualization of the data points ρ (averaged gradient approximation error) for Algorithm Setup 17 (see Appendix C.1.1) with T_d as specified applied to (4.4).

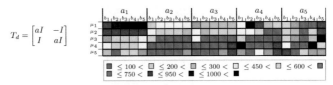

Figure C.15. Visualization of the data points k^* (steps to converge into $\mathcal{U}_{\sqrt{h}}(x^*)$) for Algorithm Setup 18 (see Appendix C.1.1) with T_d as specified applied to (4.4).

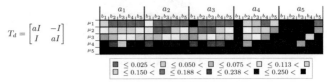

Figure C.16. Visualization of the data points ρ (averaged gradient approximation error) for Algorithm Setup 18 (see Appendix C.1.1) with T_d as specified applied to (4.4).

$$T_d = \begin{bmatrix} aI & -I \\ I & aI \end{bmatrix}$$

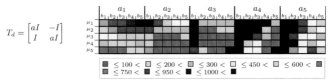

Figure C.17. Visualization of the data points k^* (steps to converge into $\mathcal{U}_{\sqrt{h}}(x^*)$) for Algorithm Setup 19 (see Appendix C.1.1) with T_d as specified applied to (4.4).

$$T_d = \begin{bmatrix} aI & -I \\ I & aI \end{bmatrix}$$

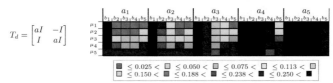

Figure C.18. Visualization of the data points ρ (averaged gradient approximation error) for Algorithm Setup 19 (see Appendix C.1.1) with T_d as specified applied to (4.4).

C.2 Benchmarking

C.2.1 Algorithm Setups

In the following the algorithm setups Algorithm Setup 20 – Algorithm Setup 27 for the benchmarking and performance evaluation in Section 4.5 are presented (four settings of McOpt, two other algorithms of the gradient approximation class, one local direct search algorithm, and the random search algorithm). The number of iterations is chosen to $k_{max} = 5000/T$, where T is the number of function evaluations per iterations. For every dimension n a set of 20 initial values x_0 is generated with $\|x_0\|_2 \leq n$.

For the sake of completeness, we state the comparison algorithms Simultaneous Perturbation Stochastic Approximation (SPSA), Finite Differences Stochastic Approximation (FDSA), Gradientless Descent (GLD), and Random Search (RS). They are presented at the end of this subsection as Algorithm 2 – Algorithm 5. For a detailed description we refer to Spall (1992), Blum (1954), Golovin et al. (2019), and Rastrigin (1963), respectively.

Algorithm Setup 20. McOpt with characteristic **[C2]** corresponding to (3.33) with

• Generating functions: $f(z) = \sqrt{z}\sin(10\ln(z))$, $g(z) = \sqrt{z}\cos(10\ln(z))$

• $T_d = \begin{bmatrix} 0 & -I \\ I & 0 \end{bmatrix} \in \mathbb{R}^{2n \times 2n}$ and singular values of W are set to $\sigma = \{1, \ldots, 1\}$

• Decreasing step size $h_k = h_0(2\lfloor k/m \rfloor + 1)^{-1}$ with $h_0 = 1$ (see Section 4.4)

and a randomized restarting of the exploration sequence is applied (see Section 4.4). The exploration sequence length is $m = 14, 28, 68$ for $n = 2, 4, 10$, respectively.　　　　●

Algorithm Setup 21. McOpt with characteristic **[C1]** corresponding to (3.40) with

• Generating functions: $f(z) = \sqrt{z}\sin(9.5\ln(z))$, $g(z) = \sqrt{z}\cos(9.5\ln(z))$

• $T_d \in \begin{bmatrix} I & -I \\ I & I \end{bmatrix} \mathbb{R}^{2n \times 2n}$ and singular values of W are (calculated to) $\sigma = \sqrt{2}\{1, \ldots, 1\}$

• Decreasing step size $h_k = h_0(2\lfloor k/m \rfloor + 1)^{-1}$ with $h_0 = 1$ (see Section 4.4)

and a randomized restarting of the exploration sequence is applied (see Section 4.4). The exploration sequence length is $m = 8, 17, 40$ for $n = 2, 4, 10$, respectively.　　　　●

Algorithm Setup 22. McOpt with characteristic **[C2]** corresponding to (3.34) with

• Generating functions: $f(z) = \sin(z)$, $g(z) = \cos(z)$

• $T_d \in \begin{bmatrix} Q & -I \\ I & Q \end{bmatrix} \mathbb{R}^{2n \times 2n}$ with Q and singular values of W as in Corollary 3

• Decreasing step size $h_k = h_0(2\lfloor k/m \rfloor + 1)^{-1}$ with $h_0 = 1$ (see Section 4.4) and a randomized restarting of the exploration sequence is applied (see Section 4.4). The exploration sequence length is $m = 3, 5, 11$ for $n = 2, 4, 10$, respectively. •

Algorithm Setup 23. McOpt with characteristic **[C1]** corresponding to (3.41) with

- Generating functions: $f(z) = \sin(z)$, $g(z) = \cos(z)$

- $T_d \in \begin{bmatrix} Q & -I \\ I & Q \end{bmatrix} \mathbb{R}^{2n \times 2n}$ with Q as in Corollary 2 and singular values of W are set to $\sigma = \{1, \ldots, 1\}$

- Decreasing step size $h_k = h_0(2\lfloor k/m \rfloor + 1)^{-1}$ with $h_0 = 1$ (see Section 4.4)

and a randomized restarting of the exploration sequence is applied (see Section 4.4). The exploration sequence length is $m = 14, 28, 68$ for $n = 2, 4, 10$, respectively. •

Algorithm Setup 24. Simultaneous perturbation stochastic approximation algorithm as in Algorithm 2 (p. 107) with initial step size $h_0 = 0.1$, and decreasing factors $\alpha = 0.602$ and $\gamma = 0.101$ (values are taken from Spall (1998)). •

Algorithm Setup 25. Finite difference stochastic approximation algorithm as in Algorithm 3 (p. 108) with initial step size $h_0 = 0.1$, and decreasing factors $\alpha = 0.3$ and $\gamma = 0.3$. •

Algorithm Setup 26. Gradientless descent as in Algorithm 4 (p. 108) with standard Gaussian sampling distribution $\mathcal{D} = \mathcal{N}(0, I)$, $R = n$, and $r = 10^{-5}n^{-1/2}$. •

Algorithm Setup 27. Random search algorithm as in Algorithm 5 (p. 108) with $d_{\min} = -5$ and $d_{\max} = 5$. •

Comparison Algorithms

Algorithm 2 Simultaneous Perturbation Stochastic Approximation (SPSA)

1: **Input:** $J : \mathbb{R}^n \to \mathbb{R}$: objective function; $k_{\max}$: number of iterations; x_0: initial value; h_0 : initial step size; α, γ: decreasing factors; Δ: sampling distribution (Bernoulli distributed with independent random variables ± 1).
2: **for** $k = 1, \ldots, k_{\max}$ **do**
3:      $h_k = h_0(k+1)^{-\alpha}$
4:      $c_k = h_0 k^{-\gamma}$
5:      Sample $\delta_k \sim \Delta$
6:      $\hat{\delta}_k = [1/(\delta_k^\top e_1) \cdots 1/(\delta_k^\top e_n)]^\top$
7:      $\hat{g}_k = \big(J(x_k + c_k\delta_k) - J(x_k - c_k\delta)\big)(2c_k)^{-1}\hat{\delta}_k$
8:      $x_{k+1} = x_k - h_k\hat{g}_k$
9: **end for**

Algorithm 3 Finite Differences Stochastic Approximation (FDSA)

1: **Input:** $J : \mathbb{R}^n \to \mathbb{R}$: objective function; $k_{\max}$: number of iterations; x_0: initial value; h_0 : initial step size; α, γ: decreasing factors.
2: **for** $k = 1, \ldots, k_{\max}$ **do**
3:     $h_k = h_0 k^{-\alpha}$
4:     $c_k = h_0 k^{-\gamma}$
5:     **for** $\ell = 1, \ldots, n$ **do**
6:         $\hat{g}_k^\top e_\ell = \big(J(x_k + c_k e_\ell) - J(x_k - c_k e_\ell) \big)(2c_k)^{-1}$
7:     **end for**
8:     $x_{k+1} = x_k - h_k \hat{g}_k$
9: **end for**

Algorithm 4 Gradientless Descent with Binary Search (GLD-Search)

1: **Input:** $J : \mathbb{R}^n \to \mathbb{R}$: objective function; $k_{\max}$: number of iterations; x_0: initial value; $\mathcal{D}$: sampling distribution (default: standard Gaussian sampling distribution $\mathcal{N}(0, I)$), R: maximum search radius, r: minimum search radius
2: $T = \log(R/r)$
3: $v_0 = 0$
4: **for** $k = 1, \ldots, k_{\max}$ **do**
5:     **for** t $= 1, \ldots$,T+1 **do**
6:         $r_t = 2^{-t} R$
7:         Sample $\tilde{v}_t \sim \mathcal{D}$
8:         $v_t = r_t \tilde{v}_t$
9:     **end for**
10:    $Y := \{ x_k + v_t | t = 0, \ldots T + 1 \}$
11:    $x_{k+1} = \arg \min_{y \in Y} \{ J(y) \}$
12: **end for**

Algorithm 5 Random Search (RS)

1: **Input:** $J : \mathbb{R}^n \to \mathbb{R}$: objective function; Δ: sampling distribution (uniform distribution with n independent random variables $\Delta_k \in [d_{\min} \, d_{\max}]$).
2: Sample $x_0 \sim \Delta$
3: **for** $k = 1, \ldots, k_{\max}$ **do**
4:     Sample $x_k \sim \Delta$
5:     $x_k = \arg\min\{ J(x_{k-1}), J(x_k) \}$
6: **end for**

C.2.2 Test Functions

The considered optimization problems for benchmarking the presented McOpt algorithm class (see Section 4.5) are given in Table C.1. Note that they are visualized for $n = 2$ in Figure 4.17 in Section 4.5.

Name	$J(x)$	$x^* \in \mathbb{R}^n$	$J(x^*)$
Sphere	$\sum_{i=1}^{n} x_{[i]}^2$	0	0
Branin	$\sum_{i=1}^{n/2} \left[(x_{[2i]} - 5.1 x_{[2i-1]}^2 + 5\pi^{-1} x_{[2i-1]} - 6)^2 \right]$ $+ 10(1 - (8\pi)^{-1}) \cos(x_{[2i]}) + 10$	$\begin{bmatrix} x_{[2i-1]}^* \\ x_{[2i]}^* \end{bmatrix} \in \left\{ \begin{bmatrix} -\pi \\ 12.275 \end{bmatrix}, \begin{bmatrix} \pi \\ 2.275 \end{bmatrix}, \begin{bmatrix} 9.424 \\ 2.475 \end{bmatrix} \right\}$	0.4063
Ellipsoid	$\sum_{i=1}^{n} 10^{2(i-1)/(n-1)} x_{[i]}^2$	0	0
Styblinski	$\sum_{i=1}^{n} (x_{[i]}^4 - 16 x_{[i]}^2 + 5 x_{[i]} + 78.334)$	$x_{[i]}^* = \pm 2.9035$	0
Rosenbrock	$\sum_{i=1}^{n-1} 100(x_{[i+1]} - x_{[i]}^2)^2 + (1 - x_{[i]})^2$	1	0
Six Hump	$\sum_{i=1}^{n/2} \left[(4 - 2.1 x_{[2i-1]}^2 + 1/3 x_{[2i-1]}^4) x_{[2i-1]}^2 \right.$	$\begin{bmatrix} x_{[2i-1]}^* \\ x_{[2i]}^* \end{bmatrix} \in \left\{ \begin{bmatrix} 0.0898 \\ -0.7126 \end{bmatrix}, \begin{bmatrix} -0.0898 \\ 0.7126 \end{bmatrix} \right\}$	0
Camel	$\left. (4 x_{[2i]}^2 - 4) x_{[2i]}^2 + 1.0317 \right]$		
Schwefel	$\sum_{i=1}^{n} \left(\sum_{j=1}^{i} x_{[j]} \right)^2$	0	0
Manevich	$\sum_{i=1}^{n} 2^{1-i} (1 - x_{[i]})^2$	1	0

Table C.1. Test functions for benchmarking the algorithm class McOpt. Note that $x_{[i]} = e_i^\top x$.

C.2.3 Logarithmic Optimality Gap Illustrations

We present in this subsection visualizations of the normalized logarithmic optimality gap as stated in (4.17) with (4.16) for the test functions given in Table C.1 for dimensions $n = 2, 4, 10$. We discuss the implications and results of Figure C.19 – Figure C.26 in Section 4.5 and do not further comment them in this subsection.

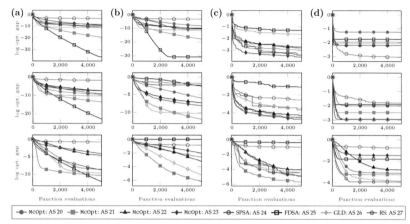

Figure C.19. Optimality gap τ for Algorithm Setup 20 – Algorithm Setup 27 (AS 20 – AS 27) as described in Appendix C.2 on the *Sphere* test function. Vertical alignment: **(Top)** $n = 2$; **(Center)** $n = 4$; **(Bottom)** $n = 10$; and horizontal alignment: **Left (a)** Normal (J); **Middle left (b)** Multiplicative noise (J_{mult}); **Middle right (c)** Additive Gaussian noise (J_{add}); **Right (d)** Discontinuous (J_{dc}).

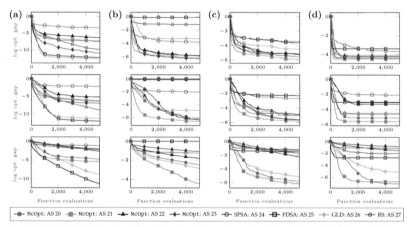

Figure C.20. Optimality gap τ for Algorithm Setup 20 – Algorithm Setup 27 (AS 20 – AS 27) as described in Appendix C.2 on the *Branin* test function. Vertical alignment: **(Top)** $n = 2$; **(Center)** $n = 4$; **(Bottom)** $n = 10$; and horizontal alignment: **Left (a)** Normal (J); **Middle left (b)** Multiplicative noise (J_{mult}); **Middle right (c)** Additive Gaussian noise (J_{add}); **Right (d)** Discontinuous (J_{dc}).

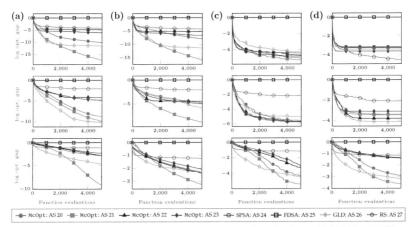

Figure C.21. Optimality gap τ for Algorithm Setup 20 – Algorithm Setup 27 (AS 20 – AS 27) as described in Appendix C.2 on the *Ellipsoid* test function. Vertical alignment: **(Top)** $n = 2$; **(Center)** $n = 4$; **(Bottom)** $n = 10$; and horizontal alignment: **Left (a)** Normal (J); **Middle left (b)** Multiplicative noise (J_{mult}); **Middle right (c)** Additive Gaussian noise (J_{add}); **Right (d)** Discontinuous (J_{dc}).

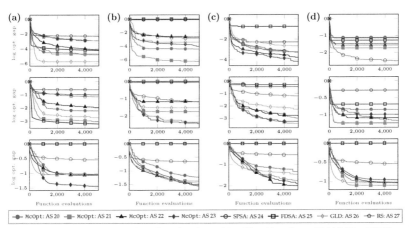

Figure C.22. Optimality gap τ for Algorithm Setup 20 – Algorithm Setup 27 (AS 20 – AS 27) as described in Appendix C.2 on the *Styblinski* test function. Vertical alignment: **(Top)** $n = 2$; **(Center)** $n = 4$; **(Bottom)** $n = 10$; and horizontal alignment: **Left (a)** Normal (J); **Middle left (b)** Multiplicative noise (J_{mult}); **Middle right (c)** Additive Gaussian noise (J_{add}); **Right (d)** Discontinuous (J_{dc}).

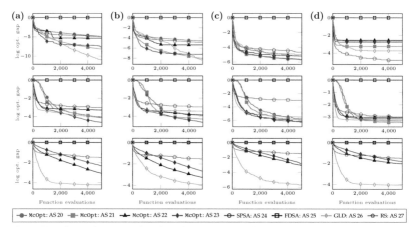

Figure C.23. Optimality gap τ for Algorithm Setup 20 – Algorithm Setup 27 (AS 20 – AS 27) as described in Appendix C.2 on the *Rosenbrock* test function. Vertical alignment: **(Top)** $n = 2$; **(Center)** $n = 4$; **(Bottom)** $n = 10$; and horizontal alignment: **Left (a)** Normal (J); **Middle left (b)** Multiplicative noise (J_{mult}); **Middle right (c)** Additive Gaussian noise (J_{add}); **Right (d)** Discontinuous (J_{dc}).

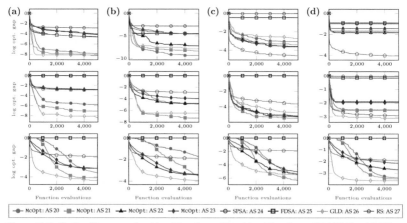

Figure C.24. Optimality gap τ for Algorithm Setup 20 – Algorithm Setup 27 (AS 20 – AS 27) as described in Appendix C.2 on the *SixHumpCamel* test function. Vertical alignment: **(Top)** $n = 2$; **(Center)** $n = 4$; **(Bottom)** $n = 10$; and horizontal alignment: **Left (a)** Normal (J); **Middle left (b)** Multiplicative noise (J_{mult}); **Middle right (c)** Additive Gaussian noise (J_{add}); **Right (d)** Discontinuous (J_{dc}).

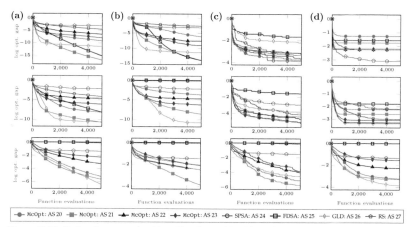

Figure C.25. Optimality gap τ for Algorithm Setup 20 – Algorithm Setup 27 (AS 20 – AS 27) as described in Appendix C.2 on the *Schwefel* test function. Vertical alignment: **(Top)** $n = 2$; **(Center)** $n = 4$; **(Bottom)** $n = 10$; and horizontal alignment: **Left (a)** Normal (J); **Middle left (b)** Multiplicative noise (J_{mult}); **Middle right (c)** Additive Gaussian noise (J_{add}); **Right (d)** Discontinuous (J_{dc}).

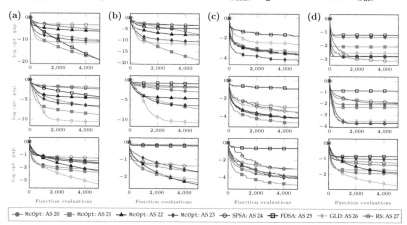

Figure C.26. Optimality gap τ for Algorithm Setup 20 – Algorithm Setup 27 (AS 20 – AS 27) as described in Appendix C.2 on the *Manevich* test function. Vertical alignment: **(Top)** $n = 2$; **(Center)** $n = 4$; **(Bottom)** $n = 10$; and horizontal alignment: **Left (a)** Normal (J); **Middle left (b)** Multiplicative noise (J_{mult}); **Middle right (c)** Additive Gaussian noise (J_{add}); **Right (d)** Discontinuous (J_{dc}).

D

Construction of Exploration Sequence Matrix W

In this section we summarize the main steps to construct an exploration sequence matrix W based on the constructive proof of Theorem 4 in Appendix B.5. A MATLAB toolbox of the described construction procedure in the sequel can be found here: https://arxiv.org/src/2006.00801v1/anc.

D.1 Step-by-step Construction of W

The exploration sequence matrix $W \in \mathbb{R}^{2n \times m}$ (n is the dimension of the optimization variable, and m the exploration sequence length) is constructed by a singular-value decomposition

$$W = U \Sigma V^\top \tag{D.1}$$

with $U \in \mathbb{R}^{2n \times 2n}$, $\Sigma \in \mathbb{R}^{2n \times m}$, and $V \in \mathbb{R}^{m \times m}$, where m has to be determined. In the following we construct U, Σ, V by

- $U = [a_1 \, b_1 \, \cdots a_n \, b_n]$, $\tag{D.2}$

- $\Sigma = \begin{bmatrix} \Sigma_0 & 0 \\ 0 & 0 \end{bmatrix}$, $\Sigma_0 = \mathrm{diag}([\sigma_1 \cdots \sigma_r])$, $\tag{D.3}$

- $V = \begin{bmatrix} \Theta - \epsilon \mathbb{1} \mathbb{1}^\top \Theta & m^{-1/2} \mathbb{1} \\ -\mathbb{1}^\top \Theta + \epsilon(m-1)\mathbb{1}^\top \Theta & m^{-1/2} \end{bmatrix}$ with $\epsilon = (m-1)^{-1}(1 - m^{-1/2})$ and

  $\Theta \in \mathbb{R}^{(m-1) \times (m-1)}$ s.t. $\Theta^\top \Theta = \Theta \Theta^\top = I$, $\tag{D.4}$

as already stated in (B.13) to (B.15). The involved components are determined in the next step-by-step construction.

Step 1: *Choose design parameters.* Select the map parameters $\alpha_1, \alpha_2 \in \mathbb{R}$ with $\alpha_1 + \alpha_2 \neq 0$, and $T_d \in \mathbb{R}^{2n \times 2n}$ according to Theorem 5 and Theorem 6 (and define $r := \mathrm{rk}(T_d)$).

Step 2: *Calculate eigenvalues and eigenvectors of* T_d. Calculate eigenvalues of T_d with $\gamma_\ell \pm \delta_\ell i$ where $\gamma_\ell \in \mathbb{R}$, $\delta_\ell \in \mathbb{R}_{\geq 0}$ and eigenvectors of T_d with $a_\ell \pm b_\ell i$ where $a_\ell, b_\ell \in \mathbb{R}^{2n}$ for $\ell = 1, \ldots, n$. The eigenvalues are sorted according to $\delta_1 \geq \delta_2 \geq \cdots \geq \delta_n \geq 0$. Note that $\gamma_\ell = 0$ if T_d skew-symmetric and $\gamma_p = 0$, $\delta_p = 0$ for $p = \lceil r/2 \rceil + 1, \ldots, n$.

Step 3: *Calculate* U. Construct as specified in (D.2), based on eigenvectors of T_d (determined in Step 2).

Step 4: *Design* Σ. The principal submatrix of Σ is constructed as $\Sigma_0 = \mathrm{diag}([\sigma_1, \ldots, \sigma_r]) \in \mathbb{R}^{r \times r}$ with the singular values σ_k of W. This step has various degrees of freedom to influence the sequence length m; distinguished in the following:

I. T_d skew-symmetric

    i) $m = r + 1$

    ii) $m \geq r + 1$ (m determined in Step 5)

II. T_d as in (3.54) in Corollary 2

    i) $m = 2n + 1$

    ii) $m \geq 2n + 1$ (m determined in Step 5)

III. T_d normal and $(2\alpha_2 - (\alpha_1 + \alpha_2)^2)(T_d + T_d^\top)$ positive definite (m determined in Step 5)

For I.i) and II.i), calculate the eigenvalues of $\tilde{P}(m)$ in (3.26) with $\mu \pm \omega_k i$ for $k = 1, \ldots, \lceil m/2 \rceil$, $\mu \in \mathbb{R}, \omega_k \in \mathbb{R}_{\geq 0}$ and $\omega_1 \geq \omega_2 \geq \cdots \geq \omega_{\lceil m/2 \rceil} \geq 0$. The choices of $\sigma_{2\ell-1}, \sigma_{2\ell}$ for $\ell = 1, \ldots, \lceil r/2 \rceil$ for each case above is presented in the following:

I.i) $\sigma_{2\ell} = \delta_\ell \omega_\ell^{-1} \sigma_{2\ell-1}^{-1}$ and $\sigma_{2\ell-1} \in \mathbb{R}_{>0}$

I.ii) $\sigma_q \in \mathbb{R}_+$, $q = 1, \ldots, r$

II.i) $\sigma_{2\ell-1} = \sigma_{2\ell} = \omega_\ell^{-1/2}$

II.ii) $\sigma_{2\ell-1} = \sigma_{2\ell} \in \mathbb{R}_{>0}$

III. $\sigma_{2\ell-1} = \sigma_{2\ell} = \gamma_\ell^{1/2}(\alpha_2 - (\alpha_1 + \alpha_2)^2)^{-1/2}$

Note that Σ depends on m, i.e. for I.ii), II.ii), and III. m has to be determined (see Step 5) first. Then, and for cases I.i) and II.i), $\Sigma = \mathrm{diag}([\Sigma_0 \, 0_{m-r}])$ can be constructed.

Step 5: *Determine sequence length* m. If in Step 4, Σ was constructed based on I.i) or II.i), $m = r + 1$ or $m = 2n + 1$, respectively (proceed directly with Step 6). Otherwise, calculate

$$\hat{\omega}_{\pi(\ell)} = \delta_\ell \left(\sigma_{2\ell-1} \sigma_{2\ell} \right)^{-1} \tag{D.5}$$

with $\sigma_{2\ell-1}, \sigma_{2\ell}$ as designed in Step 4 for $\ell = 1, \ldots, \lfloor r/2 \rfloor$ and permutation $\pi : \{1, \ldots, \lfloor r/2 \rfloor\} \to \{1, \ldots, \lfloor r/2 \rfloor\}$ such that $\hat{\omega}_1 \geq \hat{\omega}_2 \geq \cdots \hat{\omega}_{\lfloor r/2 \rfloor} \geq 0$ hold. Construct the permutation matrix $\hat{R}(\pi) \in \mathbb{R}^{\lfloor r/2 \rfloor \times \lfloor r/2 \rfloor}$ as

$$\hat{R}(\pi) = \begin{bmatrix} e_{2\pi(1)-1} & e_{2\pi(1)} & \cdots & e_{2\pi(r)-1} & e_{2\pi(r)} \end{bmatrix}. \tag{D.6}$$

Then, set $\hat{m} = r + 1$ and apply the following steps:

(a) Calculate $\tilde{P}(\hat{m})$ as defined in (3.26).

(b) Calculate eigenvalues of $\tilde{P}(\hat{m})$ with $\mu \pm \omega_k i$ where $\mu \in \mathbb{R}$, $\omega_k \in \mathbb{R}_{\geq 0}$ for $k = 1, \ldots, \lceil \hat{m}/2 \rceil$ and $\omega_1 \geq \omega_2 \geq \cdots \geq \omega_{\lceil \hat{m}/2 \rceil}$.

(c) Check if the interlacing property

$$\omega_k \geq \hat{\omega}_k \geq \omega_{\lceil \hat{m}/2 \rceil - \lfloor r/2 \rfloor + k} \tag{D.7}$$

for every $k = 1, \ldots, \lfloor r/2 \rfloor$ is satisfied with $\hat{\omega}_k$ calculated in (D.5).

(d) If (c) is true, $m = \hat{m}$. Otherwise, $\hat{m} \leftarrow \hat{m} + 1$ and goto (a).

Step 6: *Calculate V.* Construct V as specified in (D.4). Therein, the required orthogonal matrix $\Theta \in \mathbb{R}^{(m-1) \times (m-1)}$ is calculated as

$$\Theta = \begin{bmatrix} \hat{R}(\pi) & 0 \\ 0 & I \end{bmatrix} \tilde{\Theta} \tag{D.8}$$

with $\tilde{\Theta} \in \mathbb{R}^{(m-1) \times (m-1)}$ based on the construction procedure for principle submatrices in Appendix D.2, where in there

$$C = \tilde{P}(m) - I\left(\alpha_2 - \frac{1}{2}(\alpha_1 + \alpha_2)^2\right) \tag{D.9}$$

with $\tilde{P}(m)$ in (3.26) and $[\hat{\omega}_k]_{k=1,\ldots,\lfloor r/2 \rfloor}$ calculated in (D.5) has to be applied.

Step 7: *Determine W.* Finally, construct W according to (D.1).

D.2 Construction Procedure of a Principal Submatrix

In this section we present a procedure to construct an orthogonal matrix $\Theta \in \mathbb{R}^{p \times p}$, such that for a given skew-symmetric matrix $C \in \mathbb{R}^{p \times p}$ with eigenvalues $\pm \eta_\ell i$, $\eta_\ell \in \mathbb{R}_{\geq 0}$ for $\ell = 1, \ldots, \lfloor p/2 \rfloor$, and for given values $[\hat{\omega}_k]_{k=1,\ldots,q}$ $(p \geq 2q)$, which are satisfying the interlacing inequalities

$$\eta_k \geq \hat{\omega}_k \geq \eta_{\lceil p/2 \rceil - q + k}, \quad k = 1, \ldots, q; \tag{D.10}$$

it holds that

$$\Theta^\top C \Theta = \begin{bmatrix} Q & * \\ * & * \end{bmatrix} \quad \text{with}$$

$$Q = \text{diag}([Q_1 \, Q_2 \ldots Q_q]), \quad Q_k = \begin{bmatrix} 0 & -\hat{\omega}_k \\ \hat{\omega}_k & 0 \end{bmatrix}.$$

Thus, Q is the principal submatrix of C, where q and $[\hat{\omega}_k]_{k=1,\ldots,q}$ can be chosen w.r.t. (D.10). The main procedure to construct Θ is given in Algorithm 6 (p. 120) as an iterative algorithm. In each iteration $j = 1, \ldots, \lceil p/2 \rceil - q$, a $\Theta_j \in \mathbb{R}^{p \times p}$ is constructed, described in the sub-routine Algorithm 8 (p. 121), such that

$$(\Theta_1 \Theta_2 \cdots \Theta_j)^\top C \Theta_1 \Theta_2 \cdots \Theta_j = \begin{bmatrix} D_j & * \\ * & * \end{bmatrix} \tag{D.11}$$

is satisfied, where $D_j \in \mathbb{R}^{2\lfloor (p-j)/2 \rfloor \times 2\lfloor (p-j)/2 \rfloor}$ is a block diagonal skew-symmetric matrix with eigenvalues $\pm \nu_k i$, $\nu_k \in \mathbb{R}_{\geq 0}$, which are satisfying the interlacing inequalities

$$\eta_k \geq \nu_k \geq \eta_{j+k}, \tag{D.12}$$

for $k = 1, \ldots, \lfloor p/2 \rfloor - j$, where D_j is determined in the sub-routine Algorithm 7 (p. 120).

In particular, the sub-routine given in Algorithm 8 (p. 121), constructs a $\hat{\Theta} \in \mathbb{R}^{r \times r}$ such that for two given skew-symmetric matrices $D_1 \in \mathbb{R}^{r \times r}$ and $D_2 \in \mathbb{R}^{s \times s}$ in block diagonal form, where $s = 2\lfloor (r-1)/2 \rfloor$ (hence s always even), it holds that

$$\hat{\Theta} D_1 \hat{\Theta}^\top = \begin{bmatrix} D_2 & * \\ * & * \end{bmatrix}, \tag{D.13}$$

and the eigenvalues $\pm \delta_k i, \delta_k \in \mathbb{R}_{\geq 0}$ for $k = 1, \ldots, \lceil r/2 \rceil$ of D_1 and the eigenvalues $\pm \zeta_k i, \zeta_k \in \mathbb{R}_{\geq 0}$ for $k = 1, \ldots, s/2$ of D_2 satisfy the interlacing inequality

$$\delta_1 \geq \zeta_1 \geq \delta_2 \geq \zeta_2 \geq \ldots \geq \zeta_{s/2} \geq \delta_{\lceil r/2 \rceil} \geq 0. \tag{D.14}$$

The sub-routine given in Algorithm 7 (p. 120) constructs a block diagonal skew-symmetric matrix $\hat{D} \in \mathbb{R}^{2(\lceil t/2 \rceil - 1) \times 2(\lceil t/2 \rceil - 1)}$ with eigenvalues $\pm \nu_j i$, $\nu_j \in \mathbb{R}_{\geq 0}$, $j = 1, \ldots, \lceil t/2 \rceil - 1$, such that for a given skew-symmetric matrix $D \in \mathbb{R}^{t \times t}$ with eigenvalues $\pm \gamma_j i$, $\gamma_j \in \mathbb{R}_{\geq 0}$, $j = 1, \ldots, \lceil t/2 \rceil$, the interlacing inequality

$$\gamma_1 \geq \nu_1 \geq \ldots \geq \nu_{\lceil t/2 \rceil - 1} \geq \gamma_{\lceil t/2 \rceil} \geq 0. \tag{D.15}$$

is satisfied. Additionally, the interlacing inequalities

$$\gamma_k \geq \hat{\omega}_k \geq \gamma_{\lceil t/2 \rceil - q + k}, \quad k = 1, \ldots, q \tag{D.16}$$

hold.

Summarizing, Algorithm 7 (p. 120) computes a principle submatrix of a dimension that is two (or one in the first iteration) less than in the previous iteration in Algorithm 6 (p. 120), while the interlacing property (D.10) is preserved by (D.16). Then, Algorithm 8 (p. 121) constructs a Θ_j (iteration $j = 1, \ldots, \lceil p/2 \rceil - q$) based on D_j, calculated in Algorithm 7 (p. 120), such that the computed principle submatrix is obtained by an orthogonal transformation as written in (D.11). This is repeated until the principal submatrix is the block diagonal matrix with eigenvalues $[\hat{\omega}_k]_{k=1,\ldots,q}$.

Note that the function eigVal$(\cdot)$ in Algorithm 7 and Algorithm 8 (p.120f) computes the eigenvalue of a matrix in decreasing order w.r.t to the imaginary part (in Algorithm 8 (p. 121) only skew-symmetric matrices are present).

Algorithm 6 Calculate Θ

1: **Given:** $C \in \mathbb{R}^{p \times p}$, $[\hat{\omega}_k]_{k=1,\dots,q}$
2: **if** $p = 2q$ **then**
3:      $[a_\ell \pm b_\ell i]_{\ell=1,\dots,p/2} = \text{eigVec}(C)$
4:      $\Theta_1 = [a_1 \ b_1 \ a_2 \ b_2 \ \cdots \ a_{p/2} \ b_{p/2}]$
5: **else** $(p > 2q)$
6:      $D_0 = C$
7:      **for** $j = 1, \dots, \lceil p/2 \rceil - q$ **do**
8:          $D_j = \texttt{calcPSMatrix}(D_{j-1}, [\hat{\omega}_k]_{k=1,\dots,q})$
9:          $\hat{\Theta}_j = \texttt{calcThetaSub}(D_{j-1}, D_j)$
10:          $\Theta_j = \begin{bmatrix} \hat{\Theta}_j & 0 \\ 0 & I \end{bmatrix} \in \mathbb{R}^{p \times p}$
11:      **end for**
12: **end if**
13: $\Theta \leftarrow \Theta_1 \Theta_2 \cdots \Theta_{\lceil p/2 \rceil - q}$

Algorithm 7 Sub-routine: calculate principal submatrix

1: **function** $\hat{D} = \texttt{calcPSMatrix}(D, [\hat{\omega}_k]_{k=1,\dots,q})$
2:      $[\pm \gamma_k i]_{k=1,\dots,\lceil t/2 \rceil} = \text{eigVal}(D))$
3:      **for** $j = 1, \dots, \lceil t/2 \rceil - 2$ **do**
4:          $\rho = \{\hat{\omega} \in [\hat{\omega}_k]_{k=1,\dots,q} : \gamma_j \leq \hat{\omega} \leq \gamma_{j+1}\}$
5:          $\nu_j = \max\{\gamma_{j+1}, \rho\}$
6:      **end for**
7:      $\nu_{\lceil t/2 \rceil - 1} = \max\left(\hat{\omega}_q, \gamma_{\lceil t/2 \rceil - 2}\right)$
8:      **for** $j = 1, \dots, \lceil t/2 \rceil - 1$ **do**
9:          $N_j = \begin{bmatrix} 0 & -\nu_j \\ \nu_j & 0 \end{bmatrix}$
10:      **end for**
11:      $\hat{D} = \text{diag}([N_1 \ N_2 \ \cdots \ N_{\lceil t/2 \rceil - 1}])$
12: **end function**

Algorithm 8 Sub-routine: calculate $\hat{\Theta}$

1: **function** $\hat{\Theta} = \texttt{calcThetaSub}(D_1, D_2)$
2: $\quad [\pm\delta_k i]_{k=1,\ldots,\lceil r/2 \rceil} = \text{eigVal}(D_1))$
3: $\quad [\pm\zeta_k i]_{k=1,\ldots,s/2} = \text{eigVal}(D_2))$
4: $\quad$ **for** $j = 1, \ldots, \lceil r/2 \rceil - 1$ **do**
5: $\qquad x_j = \left(\prod_{i=1}^{\lfloor r/2 \rfloor} (\zeta_j^2 - \delta_i^2) \right), \quad y_j = \left(2 \prod_{\substack{i=1 \\ i \neq j}}^{\lceil r/2 \rceil - 1} (\zeta_j^2 - \zeta_i^2) \right)$
6: $\quad$ **end for**
7: $\quad$ **if** r even **then**
8: $\qquad$ **for** $j = 1, \ldots, r/2 - 1$ **do**
9: $\qquad\quad$ **if** $y_j == 0$ **then**
10: $\qquad\qquad z_{2j-1} = 0$
11: $\qquad\qquad z_{2j} = 0$
12: $\qquad\quad$ **else**
13: $\qquad\qquad z_{2j-1} = (-x_j y_j^{-1} v_j^{-1})^{1/2}$
14: $\qquad\qquad z_{2j} = z_{2j-1}$
15: $\qquad\quad$ **end if**
16: $\qquad$ **end for**
17: $\qquad z_{r-1} = \left(\prod_{i=1}^{r/2} \delta_i \right) \left(\prod_{i=1}^{r/2-1} \zeta_i^{-1} \right)$
18: $\qquad z = [z_1 \, z_2 \, \cdots \, z_{r-2}]^\top$
19: $\qquad \bar{D}_1 = \begin{bmatrix} D_2 & 0 & z \\ 0 & 0 & z_{r-1} \\ -z^\top & -z_{r-1} & 0 \end{bmatrix}$
20: $\qquad [a_\ell \pm b_\ell i]_{\ell=1,\ldots,r/2} = \text{eigVec}(D_1)$
21: $\qquad [c_\ell \pm d_\ell i]_{\ell=1,\ldots,r/2} = \text{eigVec}(\bar{D}_1)$
22: $\qquad \Theta = [a_1 \, b_1 \, a_2 \, b_2 \cdots a_{r/2} \, b_{r/2}]$
23: $\qquad \bar{\Theta} = [c_1 \, d_1 \, c_2 \, d_2 \cdots c_{r/2} \, d_{r/2}]$
24: $\quad$ **else** (r odd)
25: $\qquad$ **for** $j = 1, \ldots, \lfloor r/2 \rfloor$ **do**
26: $\qquad\quad$ **if** $y_j == 0$ **then**
27: $\qquad\qquad z_{2j-1} = 0$
28: $\qquad\qquad z_{2j} = 0$
29: $\qquad\quad$ **else**
30: $\qquad\qquad z_{2j-1} = (-x_j y_j^{-1})^{1/2}$
31: $\qquad\qquad z_{2j} = z_{2j-1}$
32: $\qquad\quad$ **end if**
33: $\qquad$ **end for**
34: $\qquad z = [z_1 \, z_2 \, \cdots \, z_{r-1}]^\top$
35: $\qquad \bar{D}_1 = \begin{bmatrix} D_2 & z \\ -z^\top & 0 \end{bmatrix}$
36: $\qquad [a_\ell \pm b_\ell i]_{\ell=1,\ldots,\lceil r/2 \rceil} = \text{eigVec}(D_1)$
37: $\qquad [c_\ell \pm d_\ell i]_{\ell=1,\ldots,\lceil r/2 \rceil} = \text{eigVec}(\bar{D}_1)$
38: $\qquad \Theta = [a_1 \, b_1 \, a_2 \, b_2 \cdots a_{\lfloor r/2 \rfloor} \, b_{\lfloor r/2 \rfloor} \, a_{\lceil r/2 \rceil}]$
39: $\qquad \bar{\Theta} = [c_1 \, d_1 \, c_2 \, d_2 \cdots c_{\lfloor r/2 \rfloor} \, d_{\lfloor r/2 \rfloor} \, c_{\lceil r/2 \rceil}]$
40: $\quad$ **end if**
41: $\quad \hat{\Theta} = \Theta\bar{\Theta}$
42: **end function**

Remark 9. Note that Algorithm 8 (p. 121) is the construction procedure of the sufficient interlacing eigenvalue statement in Lemma 4, whereas Algorithm 6 (p. 120) is the construction procedure of Lemma 5.

To verify this constructive approach we follow the lines of the proof of (Marshall, Olkin, & Arnold, 1979, Lemma B.3.), in which symmetric matrices that satisfy the interlacing property (A.14) are considered. Because of minor changes in the proof, we present only the case for p odd; thus, we construct $\bar{D}_1 \in \mathbb{R}^{p \times p}$ with $D_2 \in \mathbb{R}^{(p-1) \times (p-1)}$ and $z \in \mathbb{R}^{p-1}$ as specified in Algorithm 8 (p. 121). Then

$$g(\lambda) := \det(\lambda I - \bar{D}_1) = \prod_{i=1}^{\lfloor p/2 \rfloor} (\lambda^2 + \delta_i^2), \tag{D.17}$$

$$f(\lambda) := \det(\lambda I - D_2) = \prod_{i=1}^{\lfloor p/2 \rfloor} (\lambda^2 + v_i^2)$$

$$= g(\lambda)(\lambda + z^\top (\lambda I - \bar{D}_1)^{-1} z)$$

$$= \lambda g(\lambda)\left(1 + \sum_{i=1}^{\lfloor p/2 \rfloor} \frac{z_{2i-1}^2 + z_{2i}^2}{\lambda^2 + \delta_i^2}\right) \tag{D.18}$$

hold. Let for $j = 1, \ldots, \lfloor p/2 \rfloor$

$$z_{2j-1}^2 = z_{2j}^2 = -\left(\prod_{i=1}^{\lfloor p/2 \rfloor} (\delta_j^2 - v_i^2)\right)\left(2 \prod_{\substack{i=1 \\ i \neq j}}^{\lceil p/2 \rceil - 1} (\delta_j^2 - \delta_i^2)\right)^{-1} \tag{D.19}$$

then with $\lambda = \pm v_\ell i, \ell = 1, \ldots, \lfloor p/2 \rfloor$

$$\sum_{i=1}^{\lfloor p/2 \rfloor} \frac{z_{2i-1}^2 + z_{2i}^2}{\lambda^2 + \delta_i^2} = -\sum_{i=1}^{\lfloor p/2 \rfloor} \frac{\prod_{\substack{k=1 \\ k \neq \ell}}^{\lfloor p/2 \rfloor} (\delta_i^2 - v_k^2)}{\prod_{\substack{k=1 \\ k \neq i}}^{\lfloor p/2 \rfloor} (\delta_i^2 - \delta_k^2)} = -1. \tag{D.20}$$

Thus, $f(\pm v_\ell i) = 0, \ell = 1, \ldots, \lfloor p/2 \rfloor$ and therefore $\pm v_\ell i$ are the eigenvalues of $\bar{D}_1$, while the eigenvalues of the principal submatrix D_2 are $\pm \delta_\ell i, \ell = 1, \ldots, \lfloor p/2 \rfloor$.   •

List of Symbols

The following list presents an overview of the most frequently used symbols and acronyms. Precise definitions are given at the symbol's first appearance or in Appendix A.1.

General Notation and Symbols

Sets

$\mathbb{N}$	Set of non-negative integers.
$\mathbb{N}_{>0}$	Set of positive integers.
$\mathbb{R}$	Set of reals.
$\mathbb{R}_{\geq 0}$	Set of non-negative reals.
$\mathbb{R}_{>0}$	Set of positive reals.
$\mathbb{C}$	Set of complex numbers.
$[a, b]$	Closed interval $\{x \in \mathbb{R} : a \leq x \leq b\}$.
$[a, b[$	Semi-closed interval $\{x \in \mathbb{R} : a \leq x < b\}$.
$]a, b[$	Open interval $\{x \in \mathbb{R} : a < x < b\}$.
$\mathcal{U}_\delta(x^*)$	n-ball in $\mathbb{R}^n$ with radius $\delta \in \mathbb{R}_{>0}$ and center point $x^* \in \mathbb{R}^n$.

Functions

$C^k(\mathbb{R}^n, \mathbb{R})$	k-times differentiable functions.
$\nabla J(x)$	Gradient of $J : \mathbb{R}^n \to \mathbb{R}$ with $\nabla J(x) = [\partial J/\partial x_1 \ \cdots \ \partial J/\partial x_n]^\top$

Vectors and Matrices

e_i	The i-th unit vector; vector with i-th entry being equal to one and the rest zero.
$\mathbb{1}$	All-one vector; vector with all entries being equal to one.
I	Unit matrix; matrix with diagonal elements being equal to one and the rest zero.
$(P)_{1:r}$	The $r \times r$ principle submatrix of P .
$\text{rk}(P)$	Rank of the matrix P.

Operators and Sequences

$f \circ g$	Composition of functions.
$\mathcal{O}(h)$	Term of order h
$[f,g](z)$	Lie bracket between f,g
$\lfloor x \rfloor$	Floor operator.
$\lceil x \rceil$	Ceil operator.
$k \bmod n$	Modulo operator.
$\mathbb{E}_e$	Expectation value w.r.t real random variable e.
$\{x_k\}_{k=0}^m$	Sequence $x_0, x_1, \ldots, x_m$.

Notation and Symbols Specific for McOpt

n	Dimension of the optimization problem.
$J(x)$	Objective function with $J : \mathbb{R}^n \to \mathbb{R}$.
x^*	Local minima of objective function J.
x_k	Iterations of the optimizer with $x_k \in \mathbb{R}^n$.
f, g	Gradient-generating functions with $f, g : \mathbb{R} \to \mathbb{R}$.
h	Step size.
h_k	Adaptive step size.
$s_\ell(J(x_k))$	Evaluation maps with $s_\ell : \mathbb{R} \to \mathbb{R}^n$.
$M_k^{\sqrt{h}}(x_k)$	Transition maps with $M_k^{\sqrt{h}} : \mathbb{R}^n \to \mathbb{R}^n$.
$[\alpha_1 \ \alpha_2]$	Map parameters with $\alpha_1 + \alpha_2 \neq 0$.
m	Length of exploration sequence.
$\{w_\ell\}_{\ell=0}^{m-1}$	Exploration sequence with $w_\ell \in \mathbb{R}^{2n}$.
W	Exploration sequence matrix with $W \in \mathbb{R}^{2n \times m}$.
$T(W)$	Quadratic exploration sequence terms with $T(W) \in \mathbb{R}^{2n \times 2n}$.
T_d	Desired quadratic exploration sequence terms $T_d \in \mathbb{R}^{2n \times 2n}$.
σ	Singular values of the exploration sequence matrix W with $\sigma = \{\sigma_1, \ldots, \sigma_{\mathrm{rk}(T_d)}\}$.

Acronyms

DFO	Derivative free-optimization
FDSA	Finite differences stochastic approximation
McOpt	Magic Cube Optimization algorithm
GLD	Gradientless descent
SPSA	Simultaneous perturbation stochastic approximation
RS	Random search

Bibliography

Absil, P.-A., & Kurdyka, K. (2006). On the stable equilibrium points of gradient systems. *Systems & Control Letters*, 55(7), 573–577.

Altafini, C. (2016). Nonintegrable discrete-time driftless control systems: geometric phases beyond the area rule. In *Proc.of the 55th IEEE Conf. on Decision and Control* (pp. 4692–4697).

Ariyur, K. B., & Krstic, M. (2003). *Real-time optimization by extremum-seeking control*. John Wiley & Sons.

Atta, K. T., Johansson, A., & Gustafsson, T. (2015). Extremum seeking control based on phasor estimation. *Systems & Control Letters*, 85, 37–45.

Audet, C., & Hare, W. (2017). *Derivative-free and blackbox optimization*. Springer.

Bengio, Y. (2000). Gradient-based optimization of hyperparameters. *Neural Computation*, 12(8), 1889–1900.

Benosman, M. (2016). *Learning-based adaptive control: An extremum seeking approach–theory and applications*. Butterworth-Heinemann.

Bergstra, J., & Bengio, Y. (2012). Random search for hyper-parameter optimization. *Journal of Machine Learning Research*, 13, 281–305.

Bergstra, J. S., Bardenet, R., Bengio, Y., & Kégl, B. (2011). Algorithms for hyper-parameter optimization. In *Advances in Neural Information Processing Systems* (pp. 2546–2554).

Bloch, A. M. (2003). Nonholonomic mechanics. In *Nonholonomic mechanics and control* (pp. 207–276). Springer.

Blum, J. R. (1954). Multidimensional stochastic approximation methods. *The Annals of Mathematical Statistics*, 737–744.

Bonabeau, E., Dorigo, M., Marco, D. d. R. D. F., Theraulaz, G., & Théraulaz, G. (1999). *Swarm intelligence: from natural to artificial systems* (No. 1). Oxford University Press.

Bourbaki, N. (2008). *Lie groups and Lie algebras: Chapters 1-3* (Vol. 3). Springer.

Boyd, S., & Vandenberghe, L. (2004). *Convex Optimization*. Cambridge University Press.

Braden, B. (1986). The surveyor's area formula. *The College Mathematics Journal*, 17(4), 326–337.

Brochu, E., Cora, V. M., & De Freitas, N. (2010). A tutorial on bayesian optimization of expensive cost functions, with application to active user modeling and hierarchical reinforcement learning. *arXiv preprint arXiv:1012.2599*.

Brockett, R. (2014). The early days of geometric nonlinear control. *Automatica*, 50(9), 2203–2224.

Brockett, R. W. (1976). Nonlinear systems and differential geometry. *Proc. of the IEEE*, 64(1), 61–72.

Buhmann, M. D. (2003). *Radial basis functions: theory and implementations* (Vol. 12). Cambridge University Press.

Burnett, R. (2003). Application of stochastic optimization to collision avoidance. In *Fourth International Symposium on Uncertainty Modeling and Analysis* (pp. 123–128).

Cauchy, A. (1891). Sur l'équation à l'aide de laquelle on détermine les inégalités séculaires des mouvements des planètes. *Exer. Math.*, 9, 174-195.

Cheney, E. W., & Kincaid, D. R. (2012). *Numerical mathematics and computing*. Cengage Learning.

Choi, J.-Y., Krstic, M., Ariyur, K. B., & Lee, J. S. (2002). Extremum seeking control for discrete-time systems. *IEEE Transactions on Automatic Control*, 47(2), 318–323.

Conn, A. R., Gould, N. I., & Toint, P. L. (2000). *Trust region methods* (Vol. 1). Siam.

Conn, A. R., Scheinberg, K., & Vicente, L. N. (2009). *Introduction to derivative-free optimization* (Vol. 8). Siam.

Dean, S., Mania, H., Matni, N., Recht, B., & Tu, S. (2019). On the sample complexity of the linear quadratic regulator. *Foundations of Computational Mathematics*, 1–47.

Dochain, D., Perrier, M., & Guay, M. (2011). Extremum seeking control and its application to process and reaction systems: A survey. *Mathematics and Computers in Simulation*, *82*(3), 369–380.

Dong, N., & Chen, Z. (2012a). A novel ADP based model-free predictive control. *Nonlinear Dynamics*, *69*(1-2), 89–97.

Dong, N., & Chen, Z. (2012b). A novel data based control method based upon neural network and simultaneous perturbation stochastic approximation. *Nonlinear Dynamics*, *67*(2), 957–963.

Dürr, H.-B., Stanković, M., Ebenbauer, C., & Johansson, K. H. (2013). Lie bracket approximation of extremum seeking systems. *Automatica*, *49*(6), 1538–1552.

Dürr, H.-B., Zeng, C., & Ebenbauer, C. (2013). Saddle point seeking for convex optimization problems. In *Proc. 9th IFAC Symp. Nonlinear Control Systems (NOLCOS)* (Vol. 46, pp. 540–545).

Eberhart, R., & Kennedy, J. (1995). Particle swarm optimization. In *Proc. of the IEEE International Conference on Neural Networks* (Vol. 4, pp. 1942–1948).

Fan, K., & Pall, G. (1957). Imbedding conditions for hermitian and normal matrices. *Canadian Journal of Mathematics*, *9*, 298–304.

Feiling, J., Belabbas, M.-A., & Ebenbauer, C. (2020). Gradient-approximations and multi-variable derivative-free optimization based on non-commutative maps. *Transactions on Automatic Control*, accepted.

Feiling, J., Koga, S., Krstić, M., & Oliveira, T. R. (2018). Gradient extremum seeking for static maps with actuation dynamics governed by diffusion PDEs. *Automatica*, *95*, 197–206.

Feiling, J., Labar, C., Grushkovskaya, V., Garone, E., Kinnaert, M., & Ebenbauer, C. (2019). Extremum seeking algorithms based on non-commutative maps. *IFAC-PapersOnLine*, *52*(16), 688–693.

Feiling, J., Zeller, A., & Ebenbauer, C. (2018). Derivative-free optimization algorithms based on non-commutative maps. *IEEE Control Systems Letters*, *2*(4), 743–748.

Fermi, E. (1952). *Numerical solution of a minimum problem* (Tech. Rep.). Los Alamos Scientific Lab.

Fletcher, R. (1965). Function minimization without evaluating derivatives - a review. *The Computer Journal*, *8*(1), 33–41.

Fogel, L. J., Owens, A. J., & Walsh, M. J. (1966). *Artificial intelligence through simulated evolution*. Wiley.

Frihauf, P., Krstic, M., & Başar, T. (2013). Finite-horizon LQ control for unknown discrete-time linear systems via extremum seeking. *European Journal of Control*, *19*(5), 399–407.

Gao, J. (2014). Machine learning applications for data center optimization. *Google Research*, (https://research.google/pubs/pub42542/).

Gnu general public license. (2020). Retrieved from http://www.gnu.org/licenses/gpl.html

Golovin, D., Karro, J., Kochanski, G., Lee, C., Song, X., et al. (2019). Gradientless descent: High-dimensional zeroth-order optimization. *arXiv preprint arXiv:1911.06317*.

Golovin, D., Solnik, B., Moitra, S., Kochanski, G., Karro, J., & Sculley, D. (2017). Google vizier: A service for black-box optimization. In *Proc. of the 23rd ACM SIGKDD International Conference on Knowledge Discovery and Data Mining* (pp. 1487–1495).

Grushkovskaya, V., Zuyev, A., & Ebenbauer, C. (2018). On a class of generating vector fields for the extremum seeking problem: Lie bracket approximation and stability properties. *Automatica*, *94*, 151–160.

Guay, M., & Zhang, T. (2003). Adaptive extremum seeking control of nonlinear dynamic systems with parametric uncertainties. *Automatica*, *39*(7), 1283–1293.

Hansen, N., Finck, S., Ros, R., & Auger, A. (2009). Real-parameter black-box optimization benchmarking 2009: Noiseless functions definitions. *HAL-Inra*, (https://hal.inria.fr/inria-00362633/).

Henderson, P., Islam, R., Bachman, P., Pineau, J., Precup, D., & Meger, D. (2018). Deep reinforcement learning that matters. In *Proc. 32nd AAAI Conference on Artificial Intelligence*.

Holland, J. H., et al. (1992). *Adaptation in natural and artificial systems: an introductory analysis with applications to biology, control, and artificial intelligence*. MIT Press.

Hooke, R., & Jeeves, T. A. (1961). Direct search solution of numerical and statistical problems. *Journal of the ACM (JACM)*, *8*(2), 212–229.

Horn, R. A., & Johnson, C. R. (2012). *Matrix analysis*. Cambridge University Press.

Jones, D. R., Schonlau, M., & Welch, W. J. (1998). Efficient global optimization of expensive black-box functions. *Journal of Global optimization*, *13*(4), 455–492.

Karmanov, V. (1974). Convergence estimates for iterative minimization methods. *USSR Computational Mathematics and Mathematical Physics*, *14*(1), 1–13.

Khalil, H. K. (2002). *Nonlinear systems*. Prentice Hall.

Khong, S. Z., Nešić, D., Tan, Y., & Manzie, C. (2013). Unified frameworks for sampled-data extremum seeking control: Global optimisation and multi-unit systems. *Automatica*, *49*(9), 2720–2733.

Khong, S. Z., Tan, Y., Manzie, C., & Nešić, D. (2015). Extremum seeking of dynamical systems via gradient descent and stochastic approximation methods. *Automatica*, *56*, 44–52.

Kiefer, J., Wolfowitz, J., et al. (1952). Stochastic estimation of the maximum of a regression function. *The Annals of Mathematical Statistics*, *23*(3), 462–466.

Kirkpatrick, S., Gelatt, C. D., & Vecchi, M. P. (1983). Optimization by simulated annealing. *Science*, *220*(4598), 671–680.

Klein, A., Falkner, S., Bartels, S., Hennig, P., & Hutter, F. (2016). Fast bayesian optimization of machine learning hyperparameters on large datasets. *arXiv preprint arXiv:1605.07079*.

Ko, H.-S., Lee, K. Y., & Kim, H.-C. (2008). A simultaneous perturbation stochastic approximation (spsa)-based model approximation and its application for power system stabilizers. *International Journal of Control, Automation, and Systems*, *6*(4), 506–514.

Krstić, M., & Wang, H.-H. (2000). Stability of extremum seeking feedback for general nonlinear dynamic systems. *Automatica*, *36*(4), 595–601.

Kushner, H. J., & Clark, D. S. (2012). *Stochastic approximation methods for constrained and unconstrained systems* (Vol. 26). Springer.

Labar, C., Feiling, J., & Ebenbauer, C. (2018). Gradient-based extremum seeking: Performance tuning via lie bracket approximations. In *Proc. of the European Control Conference* (pp. 2775–2780).

Leblanc, M. (1922). Sur l'electrification des chemins de fer au moyen de courants alternatifs de frequence elevee. *Revue générale de l'électricité*, *12*(8), 275–277.

Lehman, J., Chen, J., Clune, J., & Stanley, K. O. (2018). ES is more than just a traditional finite-difference approximator. In *Proc. of the Genetic and Evolutionary Computation Conference* (pp. 450–457).

Levine, S., & Koltun, V. (2013). Guided policy search. In *International Conference on Machine Learning* (pp. 1–9).

Li, L., Jamieson, K., DeSalvo, G., Rostamizadeh, A., & Talwalkar, A. (2017). Hyperband: A novel bandit-based approach to hyperparameter optimization. *The Journal of Machine Learning Research*, *18*(1), 6765–6816.

Loshchilov, I., & Hutter, F. (2016). CMA-ES for hyperparameter optimization of deep neural networks. *arXiv preprint arXiv:1604.07269*.

Lyons, T. J., & Xu, W. (2018). Inverting the signature of a path. *Journal of the European Mathematical Society*, *20*(7), 1655–1687.

Mania, H., Guy, A., & Recht, B. (2018). Simple random search provides a competitive approach to reinforcement learning. *arXiv preprint arXiv:1803.07055*.

Marshall, A. W., Olkin, I., & Arnold, B. C. (1979). *Inequalities: theory of majorization and its applications*. Springer.

Matyas, J. (1965). Random optimization. *Automation and Remote Control, 26*(2), 246–253.

Michalowsky, S., Gharesifard, B., & Ebenbauer, C. (2017). Distributed extremum seeking over directed graphs. In *Proc. of the ieee 56th conference on decision and control* (pp. 2095–2101).

Mirjalili, S. (2020). *Evolutionary machine learning techniques: Algorithms and applications*. Springer Nature.

Mirjalili, S., & Lewis, A. (2016). The whale optimization algorithm. *Advances in engineering software, 95*, 51–67.

Mirjalili, S., Mirjalili, S. M., & Lewis, A. (2014). Grey wolf optimizer. *Advances in engineering software, 69*, 46–61.

Mnih, V., Kavukcuoglu, K., Silver, D., Graves, A., Antonoglou, I., Wierstra, D., & Riedmiller, M. (2013). Playing ATARI with deep reinforcement learning. *arXiv preprint arXiv:1312.5602*.

Mnih, V., Kavukcuoglu, K., Silver, D., Rusu, A. A., Veness, J., Bellemare, M. G., ... Ostrovski, G. (2015). Human-level control through deep reinforcement learning. *Nature, 518*(7540), 529–533.

Moré, J. J., & Sorensen, D. C. (1983). Computing a trust region step. *SIAM Journal on Scientific and Statistical Computing, 4*(3), 553–572.

Moré, J. J., & Wild, S. M. (2009). Benchmarking derivative-free optimization algorithms. *SIAM Journal on Optimization, 20*(1), 172–191.

Moreau, L., & Aeyels, D. (2000). Practical stability and stabilization. *IEEE Transactions on Automatic Control, 45*(8), 1554–1558.

Nelder, J. A., & Mead, R. (1965). A simplex method for function minimization. *The computer journal, 7*(4), 308–313.

Nesterov, Y. (2003). *Introductory lectures on convex optimization: A basic course* (Vol. 87). Springer.

Nijmeijer, H., & Van der Schaft, A. (1990). *Nonlinear dynamical control systems* (Vol. 175). Springer.

Polyak, B. T. (1987). *Introduction to optimization*. Optimization Software, Inc.

Polydoros, A. S., & Nalpantidis, L. (2017). Survey of model-based reinforcement learning: Applications on robotics. *Journal of Intelligent & Robotic Systems, 86*(2), 153–173.

Poveda, J. I., & Teel, A. R. (2017). A framework for a class of hybrid extremum seeking controllers with dynamic inclusions. *Automatica, 76*, 113–126.

Powell, M. (1965). A method for minimizing a sum of squares of non-linear functions without calculating derivatives. *The Computer Journal, 7*(4), 303–307.

Powell, M. J. (2003). On trust region methods for unconstrained minimization without derivatives. *Mathematical programming, 97*(3), 605–623.

Rastrigin, L. A. (1963). The convergence of the random search method in the extremal control of many parameter system. *Automation and Remote Control, 24*(10), 1337–1342.

Recht, B. (2019). A tour of reinforcement learning: The view from continuous control. *Annual Review of Control, Robotics, and Autonomous Systems, 2*, 253–279.

Rios, L. M., & Sahinidis, N. V. (2013). Derivative-free optimization: a review of algorithms and comparison of software implementations. *Journal of Global Optimization, 56*(3), 1247–1293.

Rosenbrock, H. (1960). An automatic method for finding the greatest or least value of a function. *The Computer Journal, 3*(3), 175–184.

Rubik, E. j. (1975, January 30). *Térbeli logikaijáték*. University Budapest. (Hungarian patent 170,062)

Rudin, W. (1964). *Principles of mathematical analysis* (Vol. 3). McGraw-Hill.

Salimans, T., Ho, J., Chen, X., Sidor, S., & Sutskever, I. (2017). Evolution strategies as a scalable alternative to reinforcement learning. *arXiv preprint arXiv:1703.03864*.

Scheinker, A., & Krstić, M. (2014). Extremum seeking with bounded update rates. *Systems & Control Letters, 63*, 25–31.

Shahriari, B., Swersky, K., Wang, Z., Adams, R. P., & De Freitas, N. (2015). Taking the human out of the loop: A review of bayesian optimization. *Proc. of the IEEE*, *104*(1), 148–175.

Snoek, J., Larochelle, H., & Adams, R. P. (2012). Practical bayesian optimization of machine learning algorithms. In *Advances in Neural Information Processing Systems* (pp. 2951–2959).

Spall, J. C. (1992). Multivariate stochastic approximation using a simultaneous perturbation gradient approximation. *IEEE Transactions on Automatic Control*, *37*(3), 332–341.

Spall, J. C. (1997). A one-measurement form of simultaneous perturbation stochastic approximation. *Automatica*, *33*(1), 109–112.

Spall, J. C. (1998). Implementation of the simultaneous perturbation algorithm for stochastic optimization. *IEEE Transactions on aerospace and electronic systems*, *34*(3), 817–823.

Spall, J. C. (2005). *Introduction to stochastic search and optimization: estimation, simulation, and control* (Vol. 65). John Wiley & Sons.

Stanković, M. S., & Stipanović, D. M. (2009). Discrete time extremum seeking by autonomous vehicles in a stochastic environment. In *Proc. of the 48th IEEE Conference on Decision and Control* (pp. 4541–4546).

Sussmann, H. J., & Liu, W. (1991). Limits of highly oscillatory controls and the approximation of general paths by admissible trajectories. In *Proc. of the IEEE 30th IEEE Conference on Decision and Control* (pp. 437–442).

Sutton, R. S., & Barto, A. G. (2018). *Reinforcement learning: An introduction.* MIT Press.

Tan, Y., Moase, W. H., Manzie, C., Nešić, D., & Mareels, I. (2010). Extremum seeking from 1922 to 2010. In *Proc. of the 29th Chinese Control Conference* (pp. 14–26).

Teel, A. R., & Popovic, D. (2001). Solving smooth and nonsmooth multivariable extremum seeking problems by the methods of nonlinear programming. In *Proc. of the American Control Conference* (Vol. 3, pp. 2394–2399).

Thompson, R. (1979). Principal minors of complex symmetric and skew matrices. *Linear Algebra and its Applications*, *28*, 249 - 255.

Todorov, E., Erez, T., & Tassa, Y. (2012). Mujoco: A physics engine for model-based control. In *Proc. of the international conference on intelligent robots and systems* (pp. 5026–5033).

Van Scoy, B., Freeman, R. A., & Lynch, K. M. (2017). The fastest known globally convergent first-order method for minimizing strongly convex functions. *IEEE Control Systems Letters*, *2*(1), 49–54.

Vidyasagar, M. (2002). *Nonlinear systems analysis.* SIAM.

Watkins, C. J., & Dayan, P. (1992). Q-learning. *Machine learning*, *8*(3-4), 279–292.

Wildhagen, S., Michalowsky, S., Feiling, J., & Ebenbauer, C. (2018). Characterizing the learning dynamics in extremum seeking: The role of gradient averaging and non-convexity. In *Proc. 58th IEEE Conference on Decision and Control* (pp. 21–26).

Winfield, D. (1973). Function minimization by interpolation in a data table. *IMA Journal of Applied Mathematics*, *12*(3), 339–347.

Zhang, C., & Ordóñez, R. (2011). *Extremum-seeking control and applications: a numerical optimization-based approach.* Springer.

Zhou, K., Doyle, J. C., Glover, K., et al. (1996). *Robust and optimal control* (Vol. 40). Prentice Hall.